INTERNATIONAL SERIES OF MONOGRAPHS IN

THE SCIENCE OF THE SOLID STATE

GENERAL EDITOR: B. R. PAMPLIN

VOLUME 1

OPTICAL PROPERTIES AND BAND STRUCTURE OF SEMICONDUCTORS

OPTICAL PROPERTIES AND BAND STRUCTURE OF SEMICONDUCTORS

BY

DAVID L. GREENAWAY

AND

GÜNTHER HARBEKE
LABORATORIES R.C.A. LTD.,
ZURICH, SWITZERLAND

THE QUEEN'S AWARD
TO INDUSTRY 1966

PERGAMON PRESS

OXFORD · LONDON · EDINBURGH · NEW YORK

TORONTO · SYDNEY · PARIS · BRAUNSCHWEIG

Pergamon Press Ltd., Headington Hill Hall, Oxford
4 & 5 Fitzroy Square, London W. 1

Pergamon Press (Scotland) Ltd., 2 & 3 Teviot Place, Edinburgh 1

Pergamon Press Inc., 44–01 21st Street, Long Island City, New York 11101

Pergamon of Canada Ltd., 207 Queen's Quay West, Toronto 1

Pergamon Press (Aust.) Pty. Ltd., 19a Boundary Street, Rushcutters Bay, N.S.W. 2011, Australia

Pergamon Press S.A.R.L., 24 rue des Écoles, Paris 5e

Vieweg & Sohn GmbH, Burgplatz 1, Braunschweig

First edition 1968

Library of Congress Catalog Card No. 67–31077

PRINTED IN GREAT BRITAIN BY COMPTON PRINTING LTD.
AYLESBURY AND LONDON
08 012648 0

CONTENTS

PREFACE

THIS monograph is concerned with experimental studies of the fundamental energy band structure of semiconductors and insulators. Research in this area of solid state physics has undergone rapid expansion during recent years, and this expansion has been due to extensive work on the intrinsic optical properties, particularly the measurement of reflectance spectra, and to the emergence of conceptionally new techniques—both experimental and theoretical—for the determination of the relevant band parameters. These approaches have led to a well-established body of knowledge about the structure of energy bands in real materials. It is felt that the time is now suitable for the presentation of a more comprehensive account of this field than has so far been available. This monograph should thus be of interest not only to those who have intimate acquaintance with the specific problems of band structure determination, but to solid state physicists generally. The treatment is primarily from an experimental viewpoint, but for the sake of completeness, a number of the basic theoretical concepts—for example, the theory of interband transitions, group-theoretical classification of electronic states, and different methods of band structure calculation—are briefly dealt with.

Considerable effort has been directed to making this monograph serve as a reference work for those with interest in the electron and optical properties of semiconductors and insulators. Besides providing an overall picture of the current state of this field, detailed information is given of the available measurement techniques and results for a large number of both cubic and non-cubic materials.

ACKNOWLEDGEMENTS

WE WOULD like to record our gratitude to those who have helped, particularly in the later stages, with the production of this Monograph. In particular we thank Professor F. Bassani, Professor M. Cardona, Dr. R. Klein and Dr. W. Rehwald for their labour in critically reading the manuscript, and suggesting a number of modifications. We would also like to thank Dr. B. Seraphin for his comments on the section on electroreflectance, together with Professor H. Ehrenreich and Professor J. C. Phillips for permission to include a not inconsiderable number of their results in the text. It may be taken for granted that we are also indebted to all the people, who cannot be named separately, whose results we have quoted and reproduced in the following chapters.

CHAPTER 1

INTRODUCTION

1.1. Historical

The purpose of this monograph is to present an overall description of the techniques and results in a field that has received a great deal of attention during recent years: the study of the fundamental band structure of semiconductors. The wealth of data now available on this subject stems essentially from two factors. First, it was realized that the measurement of absorption or reflection coefficient in the visible and ultraviolet spectral region could provide much information about the position and nature of the energy bands in a large number of materials. The interpretation of much of the data is aided by the existence of definite familial relationships between materials having similar crystal structures and outer electronic configurations. The behaviour of the energy bands in isoelectronic sequences of diamond and zincblende semiconductors, and the consistent variation of energy gaps as a function of lattice parameter for the lead salts PbS, PbSe and PbTe are two examples of this.

Secondly, the last few years have witnessed the emergence of theoretical techniques for the calculation of the band structures of real solids with an accuracy commensurate with that of the available experimental data. Clearly it would be desirable to perform such a calculation by starting with an accurate knowledge of the atomic potentials and to compute the band structure entirely from first principles. A number of such calculations have been made on semiconductors in recent years but in general have not been able to predict the detailed structure of the valence and conduction bands. Thus while a first principles calculation is obviously to be preferred on fundamental grounds, for the time being more realistic approaches have found greater utility.

The great success of recent band calculations has been due to the use of pseudopotential methods in which one constructs a potential having a small number of adjustable parameters. These parameters are chosen to fit a few of the basic well-established features of the band structure (such as the optical gap or spin-orbit splitting) and the momentum versus energy curves are then calculated. This pseudopotential approach has been applied with great success to the elemental semiconductors Si and Ge, and to heavy materials such as the lead salts where relativistic corrections become extremely large.

1.2. Optical properties

In this monograph the accent is on the experimental determination of band structure; here we are concerned with the measurement and evaluation of the optical constants of semiconductors and insulators in their fundamental region, i.e. from the absorption edge out to some higher energy where no more structure is discernible in the optical constants. We concentrate on this region since the lower energy aspects of the band structure have been treated extensively by other authors. Thus we do not deal with free carrier effects and the evaluation of detailed extremal band parameters, and optical or magneto-optical effects in the region of the lowest absorption edge. The present treatment is concerned with a more general approach to the overall band structure; while our aim is primarily to investigate the higher energy bands, in many cases the results also provide a basis for a better understanding of the low energy results.

The extremely high absorption coefficients encountered throughout the fundamental absorption region make transmission measurements impractical in many cases, thus a major fraction of the results discussed in this monograph involve the analysis of normal incidence reflectance data. Such data can be processed using the Kramers–Kronig relations to yield the optical constants n and k and the real and imaginary parts of the dielectric constant ϵ_1, and ϵ_2. It is shown that for many purposes it is possible to make good energy band assignments from a direct evaluation of primary data rather than having to compute either ϵ_2 or a density of states function for each material.

1.3. Contents

The relationships between the optical constants and their evaluation from primary experimental data using the Kramers–

Kronig transforms are treated in Chapter 2. We then discuss the available instrumentation and experimental techniques for the measurement of absorption and reflectivity over the energy range from the near infrared to the vacuum ultraviolet. Chapter 4 gives a short treatment of the quantum theory of interband transitions contributing to the optical constants and an introduction to some of the group theoretical concepts which form an integral part of the band structure models dealt with in subsequent chapters. There follow two sections dealing with the experimental evaluation of band structure: the first considers cubic materials and includes the diamond and zincblende semiconductors, the lead salts and allied materials, and the alkali halides. In the second, results for anisotropic materials are discussed, in particular the hexagonal wurtzite semiconductors, where the anisotropy can be regarded as a small perturbation to a cubic system, and the rhombohedral Bi_2Te_3 type materials. In this section the effect of the selection rules for optical transitions is studied in some detail.

Four shorter chapters follow, dealing briefly with some additional methods of obtaining information about the band structure: deformation (pressure and strain) phenomena, exciton effects with some emphasis on recent results on metastable excitons, electro-optic and photoemissive effects, and valence electron plasma effects and electron energy loss phenomena. Finally a number of reference tables have been included in the text which give the relevant physical constants and detailed energy band parameters of the materials which have been discussed.

CHAPTER 2

OPTICAL CONSTANTS AND DISPERSION RELATIONS

2.1. Plane waves in isotropic media

The dispersion and absorption of a plane electromagnetic wave is described by the complex index of refraction $N = n + ik$, where n is the ordinary refractive index and k the extinction coefficient. The plane wave can be represented by

$$\mathbf{E} = \mathbf{E}_0 e^{i(\boldsymbol{\kappa}\cdot\mathbf{r}-\omega t)} \tag{2.1}$$

and

$$\mathbf{H} = \mathbf{H}_0 e^{i(\boldsymbol{\kappa}\cdot\mathbf{r}-\omega t)} \tag{2.2}$$

where $\boldsymbol{\kappa}$ is the wave propagation vector. $\boldsymbol{\kappa}$ will in general be complex. $\boldsymbol{\kappa} = \boldsymbol{\kappa}_1 + i\boldsymbol{\kappa}_2$, and the imaginary part governs the attenuation of the wave. Equations (2.1) and (2.2) form a solution of Maxwell's equations for an uncharged medium of magnetic permeability $\mu = 1$[(1)] if

$$\boldsymbol{\kappa}\cdot\boldsymbol{\kappa} = \mu_0\epsilon_0\epsilon\omega^2 = \frac{\epsilon\omega^2}{c^2}\cdot \tag{2.3}$$

Here ϵ is the complex dielectric constant which includes the effects of the displacement and conduction currents and is defined by:

$$\epsilon = \epsilon_1 + i\epsilon_2 = \epsilon_1 + i\frac{\sigma}{\omega\epsilon_0}\cdot \tag{2.4}$$

We also define the complex refractive index N by $\epsilon = N^2$, giving

$$\epsilon_1 = n^2 - k^2, \tag{2.5}$$

$$\epsilon_2 = 2nk = \frac{\sigma}{\omega\epsilon_0}\cdot \tag{2.6}$$

If κ_1 and κ_2 are parallel (homogeneous plane wave) the two fields and the direction of propagation are mutually perpendicular and we find from equation (2.3)

$$|\kappa_1| = \frac{n\omega}{c} \qquad |\kappa_2| = \frac{k\omega}{c}. \tag{2.7}$$

In this case the time-averaged energy flow is given by

$$\bar{S} = \frac{nc\epsilon_0}{2}(\mathbf{E}^* \cdot \mathbf{E})\, \kappa_E, \tag{2.8}$$

where κ_E is a unit vector in the direction of κ_1 and κ_2. With the expression for **E** from equation (2.1) we obtain

$$\bar{S} = \frac{nc\epsilon_0}{2}(\mathbf{E}_0^* \cdot \mathbf{E}_0)\kappa_E e^{-2\kappa_2 \cdot r}. \tag{2.9}$$

It is seen that the energy flow decreases by a factor $e^{-2|\kappa_2|d}$ over the distance d. The absorption coefficient K, defined by the relative decrease of energy flow per unit distance in the direction of propagation, is thus

$$K = 2|\kappa_2| = \frac{2k\omega}{c} = \frac{4\pi k}{\lambda}, \tag{2.10}$$

λ being the wavelength in vacuum.

2.2. Reflection and transmission of plane waves

2.2.1. *General formulae*

The amplitudes of reflected and refracted waves of a plane wave incident upon a plane boundary, say $z = 0$, between two media, are given by the boundary conditions requiring the continuity of the x and y components of **E** and **H**. We consider separately the cases for the components of the electric field E_n normal and E_p parallel to the plane of incidence. The boundary conditions lead to Fresnel's formulae for the amplitude reflection coefficients r_n and r_p of a wave incident at an angle ϕ from a transparent medium of refractive index n_1 on the boundary of an absorbing medium of complex refractive index N_2:

$$r_n = \frac{n_1 \cos\phi - (N_2^2 - n_1^2 \sin^2\phi)^{1/2}}{n_1 \cos\phi + (N_2^2 - n_1^2 \sin^2\phi)^{1/2}}, \tag{2.11}$$

$$r_p = \frac{N_2^2 \cos\phi - (N_2^2 n_1^2 - n_1^4 \sin^2\phi)}{N_2^2 \cos\phi + (N_2^2 n_1^2 - n_1^4 \sin^2\phi)}. \tag{2.12}$$

The reflection coefficient R is defined as the ratio of the time-averaged energy flow reflected from the surface to the incident flow. Thus $R_n = r_n^* \cdot r_n$; $R_p = r_p^* \cdot r_p$.

If the second medium is also transparent ($N_2 = n_2$) Snell's law $\sin\phi \cdot n_1 = \sin\psi \cdot n_2$, where ψ is the angle of refraction, holds and we obtain:

$$R_n = \frac{\sin^2(\psi-\phi)}{\sin^2(\psi+\phi)}, \tag{2.13}$$

$$R_p = \frac{\tan^2(\psi-\phi)}{\tan^2(\psi+\phi)}. \tag{2.14}$$

Equation (2.14) shows that R_p goes to zero if $\psi+\phi = \pi/2$. The angle of incidence is then called the Brewster angle ϕ_B and Snell's law gives $\tan\phi_B = n_2/n_1$. R_n is different from zero at the Brewster angle, hence the reflected light is totally polarized. Figure 2.1 shows R_n and R_p as a function of ϕ for $n_2 = 3$, $n_1 = 1$.

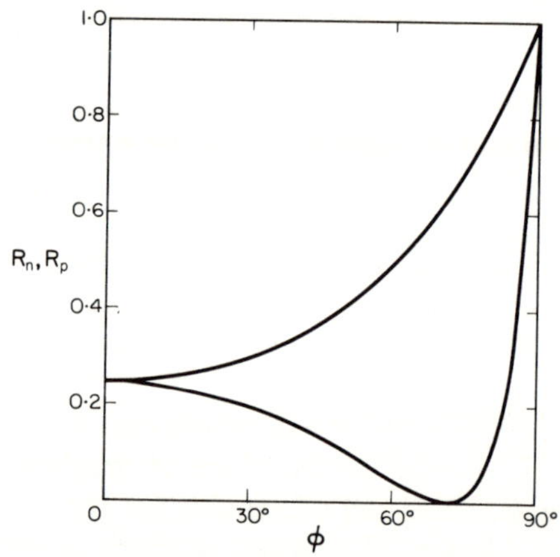

FIG. 2.1. Reflection coefficients R_n and R_p plotted against the angle of incidence for $n_2 = 3$ and $n_1 = 1$. Upper curve R_n, lower curve R_p.

It is seen that at all angles of incidence except $\phi = 0$ or $\pi/2$ $R_n > R_p$. Hence, if the incident light is unpolarized the reflected light will be either partly or totally polarized normal to the plane of incidence whereas the refracted light is always partly polarized in the plane of incidence.

If the second medium is absorbing, Snell's law is only true in a generalized form with a complex angle of refraction. This means that the planes of constant amplitude in the medium are now no longer coincident with the planes of constant phase. The refracted wave is conventionally called an inhomogeneous "plane" wave. The reflection coefficients are still given by equations (2.13) and (2.14) but ψ is complex as a consequence of the complex refractive index N_2 indicating the change of phase in the reflected wave. Hence, the reflected wave is in general elliptically polarized. There are four main groups of experimental methods which can be used in order to obtain the two optical constants simultaneously. These methods are briefly outlined below.

2.2.2. *Oblique incidence reflectivity*

If we take the first medium to be air, put $N_2 = N = n + ik$ and $N \cos \psi = a + ib$ we find $a^2 - b^2 = n^2 - k^2 - \sin^2\phi$, $ab = nk$, and for the reflected intensities:[(2)]

$$R_n = \frac{(a - \cos \phi)^2 + b^2}{(a + \cos \phi)^2 + b^2}, \tag{2.15}$$

$$R_p = R_n \frac{(a - \sin \phi \tan \phi)^2 + b^2}{(a - \sin \phi \tan \phi)^2 + b^2}. \tag{2.16}$$

By two measurements of R_n or R_p or R_p/R_n at different known angles of incidence the quantities a and b and thus n and k can be determined. In this case there is no angle at which R_p goes to zero. For small values of k, however, there is a sharp minimum in R_p/R_n close to the Brewster angle for $k = 0$. For values of k much larger than n one finds that the reflection coefficients are practically independent of ϕ and differ only by a small amount from unity. This situation applies to metals in the infrared.

2.2.3. *Azimuth and phase shift*

From equations (2.11) and (2.12) and the generalized Snell's law we find that the ratio of the complex amplitudes is given by:

$$\frac{r_n}{r_p} = -\frac{\cos (\phi - \psi)}{\cos (\phi + \psi)} = \frac{|r_n|}{|r_p|} e^{i\delta}. \tag{2.17}$$

Here we have introduced the relative phase difference δ between the two components. δ is zero for $\phi = 0$ and $\phi = \pi/4$ but otherwise nonzero. The ratio of the moduli of the amplitude coefficients is related to the azimuth ρ by $\tan \rho = |r_p|/|r_n|$. Rationalizing (2.17) shows that n and k can be found from the measurement of ρ and δ.

2.2.4. *Normal incidence reflectivity*

Equations (2.15) and (2.16) become very simple for the case of normal incidence:

$$R_n = R_p = R = \frac{(n-1)^2 + k^2}{(n+1)^2 + k^2}. \tag{2.18}$$

We see here that n and k cannot be directly determined from a reflectance measurement at normal incidence. However, the complex amplitude $r_n = r_p = r$ can be written as

$$r = \frac{n + ik - 1}{n + ik + 1} = |r|e^{i\theta}. \tag{2.19}$$

Rationalizing (2.19) gives

$$\tan \theta = 2k/(n^2 + k^2 - 1) \tag{2.20}$$

and allows n and k to be expressed in terms of $|r|$ and θ. In Section 2.3 it will be shown how θ can be found from the reflection spectrum so that n and k can also be determined from the normal incidence reflectance.

2.2.5. *Normal incidence transmission*

Of considerable practical interest is the case of transmission and reflection of plane waves by a plane parallel sheet of a medium of refractive index N in air under normal incidence. The complete formulae for the transmitted and reflected intensities I_t and I_r respectively for incident intensity $I_0 = 1$ are given by Stratton.[(3)]

$$I_t = \frac{e^{-Kd}[(1-R)^2 + 4R \sin^2 \theta]}{(1 - Re^{-Kd})^2 + 4Re^{-Kd} \sin^2(\theta + \alpha)}, \tag{2.21}$$

$$I_r = \frac{R[(1 - e^{-Kd})^2 + 4e^{-Kd} \sin^2 \alpha]}{(1 - Re^{-Kd})^2 + 4Re^{-Kd} \sin^2(\theta + \alpha)}, \tag{2.22}$$

where K, R and θ are given by equations (2.10), (2.18) and (2.20), respectively. The quantity $\alpha = 2\pi nd/\lambda$ indicates the phase change on one traverse of the sheet of thickness d. The second term in the denominator is an interference term which can be used advantageously to determine n from the interference fringes which result when K is small enough to allow the light wave to traverse the sample several times. If no interference fringes are observed because of, for example, surface roughness we have to average over the interference term and find

$$I_t = \frac{(1-R)^2 e^{-Kd}(1+k^2/n^2)}{1-R^2 e^{-2Kd}}, \tag{2.23}$$

$$I_r = \frac{R(1-e^{-2Kd})}{1-R^2 e^{-2Kd}}. \tag{2.24}$$

These equations show that n and k can be found from I_t and I_r. For most transmission experiments $k^2 \ll n^2$. If d is further chosen to ensure that $R^2 e^{-2Kd} \ll 1$ equation (2.23) takes the extremely simple form

$$I_t = (1-R)^2 e^{-Kd}. \tag{2.25}$$

Hence K can be found simply from the transmission of two samples of different thickness without knowledge of the reflectance, provided the reflectivity of both samples is the same.

2.3. Dispersion relations

Dispersion relations are of a very general nature and connect dispersive with absorptive processes. They are applicable to the real and imaginary parts of a complex quantity which describes a linear relation between two amplitudes. The only two further requirements are boundedness and causality.[4, 5] Because only these very general conditions have to be fulfilled dispersion relations are valid in many fields of physics and engineering. Originally they were derived by Kramers and Kronig[6] for the dispersion of X-rays and are therefore often called Kramers–Kronig relations. They are widely used in scattering theories and the theory of electrical circuits.[7] Here we use them for the optical region where the frequency-dependent complex quantity $\epsilon(\omega)$ expresses a linear relationship between the amplitude of the electric field $\mathbf{E}(\omega)$ and the amplitude of the displacement $\mathbf{D}(\omega)$:

$$\mathbf{D}(\omega) = \epsilon(\omega)\mathbf{E}(\omega).$$

In contrast to static fields, the complex $\epsilon(\omega)$ describes a phase shift between the two amplitudes. We can write the Kramers–Kronig relations in the form[1]

$$\epsilon_1(\omega) = 1 + \frac{2}{\pi} P \int_0^\infty \frac{\omega' \epsilon_2(\omega')}{\omega'^2 - \omega^2} d\omega', \qquad (2.26)$$

$$\epsilon_2(\omega) = -\frac{2\omega}{\pi} P \int_0^\infty \frac{\epsilon_1(\omega')}{\omega'^2 - \omega^2} d\omega', \qquad (2.27)$$

where P means the Cauchy principal value of the integral. It should be noted that this is the form for nonmetals. Because of the nonvanishing conductivity of metals for $\omega \to 0$ this case has to be treated in a modified dispersion relation. Toll[4] has shown that in the final result it leads to one additional term $\sigma_0/\epsilon_0\omega$ on the right-hand side of equation (2.27).

Dispersion relations can be given for n and k as well as $\mathrm{Re}(1/\epsilon)$ and $\mathrm{Im}(1/\epsilon)$ in a form completely analogous to equations (2.26) and (2.27) for ϵ_1 and ϵ_2. As we mentioned in Section 2.2 the most useful form of the dispersion relations for the determination of n and k from normal incidence reflection data is that which determines the phase angle θ. The complex amplitude reflection coefficient $r = |r|e^{i\theta}$ expresses a linear relationship between the amplitudes of reflected and incident light. Therefore there exists a dispersion relation connecting the real and imaginary parts of

$$\ln r = \ln |r| + i\theta = \tfrac{1}{2} \ln R + i\theta$$

given by Toll[4]:

$$\theta(\omega) = -\frac{\omega}{\pi} P \int_0^\infty \frac{\ln R(\omega')}{\omega'^2 - \omega^2} d\omega'. \qquad (2.28)$$

Thus the phase angle θ can be calculated if the function $R(\omega)$ is known over the entire spectral region and n and k can then be obtained from R and θ using equations (2.18) and (2.20).

Roessler[(8)] has shown that the Kramers–Kronig analysis can also be performed on reflection data obtained under oblique incidence. For the complex amplitude reflection coefficient r_n, for example, we have

$$r_n = |r_n| e^{i\theta} = -\frac{a+ib-\cos\phi}{a+ib+\cos\phi} \tag{2.29}$$

with a and b having the same meaning as in equation (2.15). We thus have a dispersion relation between θ and $\ln R_n$ equivalent to equation (2.28). n and k are then given by two equations in terms of a and b which in turn are given by two equations in terms of R_n and θ. The measurement with unpolarized light under an angle $\phi = 45°$, where $R_p = R_n^2$ according to equations (2.15) and (2.16), is relatively simple:

$$R_{45} = \tfrac{1}{2}(R_{n,45}+R_{p,45}) = \tfrac{1}{2}R_{n,45}(1+R_{n,45}).$$

We can remove the singularity in equation (2.28) by writing

$$\theta(\omega) = -\frac{\omega}{\pi}\int_0^\infty \frac{\ln R(\omega')-\ln R(\omega)}{\omega'^2-\omega^2}\,d\omega' \tag{2.30}$$

since

$$\int_0^\infty \frac{1}{\omega'^2-\omega^2}\,d\omega' = 0.$$

Equation (2.30) can also be written as

$$\theta(\omega) = -\frac{1}{2\pi}\int_0^\infty \ln R(\omega')\frac{d}{d\omega'}\ln\left|\frac{\omega'-\omega}{\omega'+\omega}\right| d\omega'. \tag{2.31}$$

Integrating by parts gives

$$\theta(\omega) = -\frac{1}{2\pi}\left[\ln R \ln\left|\frac{\omega'-\omega}{\omega'+\omega}\right|\right]_0^\infty + \frac{1}{2\pi}\int_0^\infty \ln\left|\frac{\omega'-\omega}{\omega'+\omega}\right|\frac{d\ln R(\omega')}{d\omega'}\,d\omega', \tag{2.32}$$

where the first term is zero. This form of the phase shift shows that a constant reflectance gives $\theta = 0$, and thus no absorption. The phase angle θ is given by the integral over the derivative of ln R times a weighting function which peaks sharply at $\omega' = \omega$ and approaches zero at frequencies far from ω.[7] Thus the derivative of ln R at the frequency ω influences the value of $\theta(\omega)$ much more than the slope of ln R at remote points on the frequency scale. This local character of the Kramer–Kronig transforms can also be seen if equations (2.26) and (2.27) are brought into a form analogous to equation (2.32). One then finds the following structural correlations first discussed by Velický[9]: ϵ_1 will have a maximum where the derivative of ϵ_2 is large and positive, i.e. at frequencies where ϵ_2 increases strongly. Correspondingly there will be a minimum in ϵ_1 for large and negative values of $d\epsilon_2/d\omega$. Maxima of ϵ_2 will occur where the slope of ϵ_1 is large and negative, i.e. at frequencies where ϵ_1 decreases sharply. For the evaluation of θ, in principle, one must know $R(\omega)$ over the entire spectral region from zero to infinity. In the case of an isolated reflectivity band where the reflectivity can be taken as constant outside the band this requirement does not cause any problem since a constant reflectivity gives no contribution to θ. In fundamental optical spectra, however, the reflectivity is mostly still varying at the highest frequencies used in the measurement. One therefore has to extrapolate the measured reflectance to higher energies in a proper way. The most reasonable extrapolation procedure assumes that above the valence electron plasma frequency ω_p the dielectric constant is given by the Drude formula:

$$\epsilon = 1 - \frac{{\omega_p}^2}{\omega^2}. \tag{2.33}$$

Thus at high frequencies $\epsilon \to 1, \epsilon_2 \to 0$ and $k \to 0$ so that

$$n = \epsilon^{1/2} \approx 1 - \frac{1}{2}\frac{{\omega_p}^2}{\omega^2} \quad \text{and} \quad R = \frac{(n-1)^2}{(n+1)^2} \approx \frac{1}{16}\frac{{\omega_p}^4}{\omega^4}.$$

If the upper frequency limit of the measurement ω_1 is larger than ω_p it seems, therefore, logical to extrapolate $R(\omega)$ beyond ω_1 by $R = R_1(\omega_1/\omega)^4$ where R_1 is the measured reflectivity at ω_1. Equation (2.32) is then written as:[10]

$$\theta(\omega) = \frac{1}{2\pi}\int_0^{\omega_1} \ln\left|\frac{\omega'-\omega}{\omega'+\omega}\right| \frac{\mathrm{d}}{\mathrm{d}\omega'}(\ln R(\omega') - \ln R(\omega))\mathrm{d}\omega'$$
$$+\frac{1}{2\pi}\ln\frac{R(\omega)}{R_1}\ln\left|\frac{\omega_1+\omega}{\omega_1-\omega}\right|$$
$$+\frac{4}{\pi}\sum_{n=0}^{\infty}\frac{1}{(2n+1)^2}\left(\frac{\omega}{\omega_1}\right)^{2n+1}[1+(2n+1)\ln\omega_1]. \quad (2.34)$$

Another approach which has been taken[11] is to extrapolate R in the form:

$$R = R_1\left(\frac{\omega_1}{\omega}\right)^p,$$

where the exponent p is fitted in order to give the best agreement with the experiment at low energies. A certain criterion for a good choice of p is the condition that the phase angle below the absorption edge has to be zero although it should be borne in mind that the weighting function tends to small values for $\omega < \omega_{\mathrm{edge}}$ and large ω_1. A third method of extrapolating R is to assume $R = R_1 e^{B(\omega_1-\omega)}$, where B is taken to fit the measured derivative of R at ω_1. Here the integration range must be cut off at a certain frequency ω_2 in order to avoid divergence of the integral. ω_2 is again chosen to fit the low energy data.

2.4. Sum rules

For high frequencies $\omega \gg \omega_u$, where ω_u is a certain limiting frequency such that there is no absorption above ω_u, one can set the upper limit of the integral in equation (2.26) at ω_u and because of $\omega' \leqslant \omega_u \ll \omega$ one obtains

$$\epsilon_1(\omega) = 1 - \frac{2}{\pi\omega^2}\int_0^{\omega_u} \omega' \epsilon_2(\omega')\mathrm{d}\omega'.$$

The Drude formula (2.33) valid in this region then leads to the sum rule:

$$\int_0^{\omega_u} \omega' \epsilon_2(\omega')\mathrm{d}\omega' = \frac{\pi}{2}\omega_p{}^2 = \frac{\pi N_v e^2}{2m\epsilon_0} \quad (2.35)$$

where m is the free electron mass and N_v the valence electron density. This formula provides a definite check on the presence of further absorption bands above ω_u. If the integral is evaluated out to a frequency ω_0:

$$\int_0^{\omega_0} \omega' \epsilon_2(\omega')\,d\omega' = \frac{\pi e^2}{2m\epsilon_0} N_{\text{eff}} = \frac{\pi e^2 N_{\text{at}}}{2m\epsilon_0} n_{\text{eff}} \tag{2.36}$$

where N_{at} is the atom density, one can regard[12] n_{eff} as the number of valence electrons per atom contributing to the optical constants in the frequency interval from zero to ω. Another sum rule is obtained for $\omega \rightarrow 0$:

$$\epsilon(0) \equiv \epsilon_1(0) = 1 + \frac{2}{\pi} \int_0^{\infty} \frac{\epsilon_2(\omega')}{\omega'}\,d\omega'. \tag{2.37}$$

This relation illustrates that the static dielectric constant $\epsilon(0)$ is determined by ϵ_2 (ω) over the entire spectral region. Because of the denominator particularly high values of $\epsilon(0)$ arise if ϵ_2 possesses high values at low frequencies. If we restrict the discussion to interband transitions it can be inferred that materials with low energy gaps and strong interband transitions at energies just above the gap are likely to have high values of the static dielectric constant. Analogous to equation (2.36) we can define an effective static dielectric constant

$$\epsilon_{\text{eff}}(0) \equiv \epsilon_{1,\text{eff}}(0) = 1 + \frac{2}{\pi} \int_0^{\omega_0} \frac{\epsilon_2(\omega')}{\omega'}\,d\omega'. \tag{2.38}$$

A further sum rule exists for $\text{Im}(-1/\epsilon)$ which is equivalent to equation (2.35)

$$\int_0^{\omega_u} \omega' \text{Im}\left(-\frac{1}{\epsilon}\right) d\omega' = \frac{\pi N_v e^2}{2m\epsilon_0}. \tag{2.39}$$

CHAPTER 3

EXPERIMENTAL METHODS

3.1 Introduction

In this chapter we are concerned with the instrumentation and techniques available for the measurement of absorption coefficient, absolute reflection coefficient, and other optical properties of semiconductors (and insulators) primarily in their fundamental region, i.e. in the region near and above the lowest forbidden energy gap. This region extends from the near infrared to the vacuum ultraviolet. Fortunately, strong interband transitions are exhausted in all but the most ionic of the insulators at 25 eV, the practical limit of the measurements and techniques we describe; thus we do not attempt to go beyond this point. We lay some emphasis on instrumentation in the vacuum ultraviolet, as up to now there is relatively little information available in the literature about low resolution absolute measurements in this region. We detail some of the methods available for measurement of reflectivity at reduced temperatures as again this is a somewhat undocumented field. Brief mention is also made of techniques for pressure and strain effect determination, measurement of electroreflectance, and the measurement of photoemission from semiconductors.

3.2. Equipment

3.2.1. *Spectrometers and monochromators*

A catalogue of the wide range of instruments currently available is beyond the scope of this monograph. It is worth while, however, to point out a few of the considerations which can determine the choice of an instrument for the measurement of optical spectra of solids. Many spectrometers, particularly double-beam instruments, are designed primarily for chemical analysis. Such instruments may not possess a beam focus in the sample position and consequently suffer a large loss in sensitivity if one attempts to measure small area samples. The compact nature of many semiautomatic spectrometers makes it often difficult to find space without alteration to install auxiliary equipment such as a cryostat or reflectometer. Normal incidence-reflectance attachments are available for most double-beam spectrometers (consisting usually of a reflecting prism or two-mirror arrangement which is inserted into the sample beam) but these devices cannot in general be used to measure absolute reflectivity unless a calibration curve is obtained for the reflectance of the mirror surface in the attachment. This is possible in the infrared, where aluminium has a constant high reflectivity, but such calibrations become unreliable as one approaches the ultraviolet end of the spectrum.

The best solution (certainly where cost is a consideration) is probably a single-beam instrument with an additional optical system for both absorption and reflectance measurements (see Section 3.3).Prism monochromators do not have the associated problem of filtering out higher-order spectra, but grating instruments usually have higher resolution and are much easier to calibrate. In the vacuum ultraviolet gratings are now used almost exclusively. Double-pass instruments – which must be used in a.c. operation – are particularly useful in the infrared as they generally combine reasonably high resolving power with low scattered light. Below 2000 Å (which we define as the onset of the vacuum ultraviolet region) the reflectivity of aluminium deteriorates rapidly with age[13] and it is desirable to use spectrometers which have a minimum number of reflecting surfaces. Although recent advances in the technique of protecting mirror surfaces with thin layers of LiF[14] and MgF_2[15] can ensure high reflectivity in the 1200–2000 Å region, the method is not applicable to shorter wavelengths. Rowland circle-type spectrometers provide the necessary optical simplicity in that they use only one reflecting element – a concave grating – but such instruments are mechanically complex if used for photoelectric scanning.

Figure 3.1 shows schematically another spectrometer suitable for the vacuum ultraviolet, the Seya–Namioka type; we illustrate it here as it is an elegantly simple design which has been used extensively in recent years for solid state measurements. The principle is described fully in the original papers by Seya[16] and

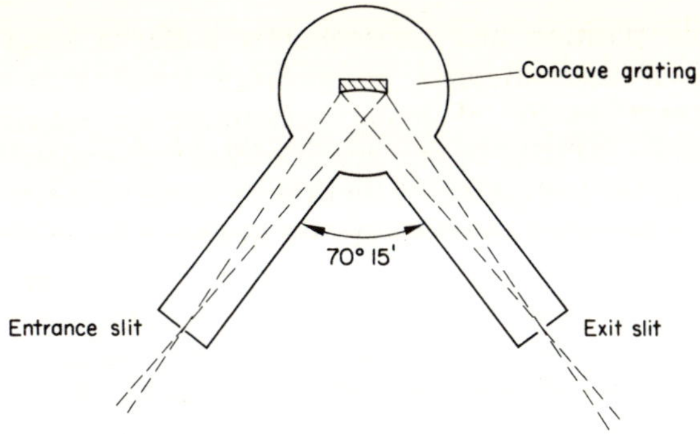

FIG. 3.1. Seya–Namioka system.

Namioka[17, 18]. Both the angle and the distance between the entrance and the exit slits and the grating remain constant as the grating is rotated, yet focus is preserved throughout the vacuum ultraviolet and a resolution of better than 1 Å can be obtained with a 1-metre focal length grating. This resolution limit is determined by the astigmatism in the Seya instrument which is severe and which also causes some loss in energy over normal incidence spectrometers. These drawbacks, however, are more than offset by the extreme simplicity of construction and operation, and the fact that a great deal of space is available around the slits for installing source and detection equipment.

3.2.2. *Sources*

The infrared and near infrared regions present no problems: the traditional Globar and Nernst filament provide strong stable continua. Both sources are operated in air and have good lifetimes. The Nernst filament can be operated at a higher temperature than the Globar and thus gives better performance in the near infrared. Beyond a wavelength of about 30μ, however, the superior emissivity of the Globar makes it the more desirable of the two. Tungsten filament lamps with either glass or quartz envelopes provide adequate sources for the near infrared and visible. Two other convenient sources for more intense visible radiation are now available: the high pressure xenon arcs (such as those manufactured by Osram in Europe) and the quartz iodine lamps. Both emit essentially a continuous spectrum and in addition the former also provides an extremely intense source of near ultraviolet radiation. Xenon arc lamps are also produced by Hanovia with high u.v. transmitting quartz envelopes which are useful down to 1600 Å.

Vacuum ultraviolet sources fall into two classes: continuous and line sources: the latter provide the greater intensity. Over the past few years much work has gone into the production of more intense vacuum ultraviolet sources, but the d.c. discharge through low-pressure hydrogen continues to be the most useful source for $\lambda > 900$ Å; this has a continuum extending from the visible down to ~1650 Å and below that a complex line spectrum. In addition to its strong emission line at 584 Å helium possesses two continua 4000–1050 Å[19] and 1000–600 Å[20]; they can best be excited using a disruptive discharge. The intensities of those continua increase as the gas pressure is increased and for the higher energy continuum at least, differential pumping at the spectrometer entrance slit may be necessary. Other rare gases (e.g. Ar, Xe, Kr) produce continua in the 1000–2200 Å region. These gases can be excited normally[21] or by using microwave power.[22] Spark

TABLE 3.1

Vacuum ultraviolet sources

All the sources shown here are operable under a wide range of both pressure and excitation conditions. The numbers given are intended only as a guide to typical conditions.

Source	Emission and range	Excitation
High-pressure Xe arc	Continuum with superimposed lines 10,000–2000 Å	d.c. 50 V 10 A
Low-pressure D_2 arc	Continuum: 6000–2000 Å	d.c. 200 V 250 mA
Low-pressure H_2 discharge (0·1–4 mm)	Continuum: 6000–1650 Å Line spectrum 1650–900 Å	d.c. 10 kV 50 mA
High-pressure He discharge (50–500 mm)	Continuum: 4000–1050 Å 1000–600 Å	a.c. disruptive discharge in kilocycle range
Low-pressure He discharge	Emission line at 584 Å	d.c. 10 kV 50 mA
High-pressure Xe discharge High-pressure Kr discharge High-pressure Ar discharge	Continuum 2200–1500 Å Continuum 1800–1250 Å Continuum 1600–1090 Å	a.c. disruptive discharge or microwave excitation 200 watts, 2450 MHz
Electron synchrotron	Continuum, visible – 60 Å (polarized radiation)	Peak emission depends on beam energy, e.g. 300 MeV gives max. at 85 Å

sources having good intensity are also available, but generally the repetition rate is low, making photoelectric recording difficult. Hunter[(23)] has described a low-pressure spark source in which the discharge is controlled with a thyratron. Stable operation is possible with repetition rates of up to 80/sec. A number of convenient sources for the ultraviolet and vacuum ultraviolet regions are shown in Table 3.1; the excitation mode for each is also given.

3.2.3. *Detectors (vacuum ultraviolet)*

Reviews of currently available detectors for the vacuum ultraviolet have been given by Tousey[(24)] and Dunkelman *et al.*[(25)] Here we shall only mention two detectors which have been found convenient for the type of measurement described in this monograph.

Undoubtedly the simplest to use (and probably the most sensitive) is the 1P21 photomultiplier used in conjunction with a phosphor. Phosphors such as sodium salicylate or terphenyl combine good sensitivity with approximately constant quantum efficiency over an extremely wide energy range and in fact these 1P21–phosphor combinations cover the range from ~ 6000 Å to below 500 Å. In practice a thin quartz plate coated with a thin (0·1 mm) layer of phosphor is placed directly in front of the multiplier. These layers can be prepared by spraying a dilute solution in an organic solvent (e.g. alcohol) on to hot (150°C) quartz plates. For maximum efficiency the layers should consist of very finely divided material. The main drawback of the 1P21–phosphor combinations is their sensitivity to long wavelength scattered radiation, but this can be largely overcome if an a.c. detection system is employed and the radiation modulated inside the vacuum housing of the spectrometer using a chopper which is transparent to the unwanted long wavelengths (for example, a thin quartz chopper for measurements below 1500 Å and a LiF chopper below 1050 Å). Phase-sensitive detection systems can operate at a rather low frequency (~ 15 Hz), thus both rotary and magnetic drives through the vacuum wall of the spectrometer are completely feasible.

Stray light effects can also be minimized, both in a.c. and d.c. detection systems, by using solar blind multipliers. Such tubes are now available commercially (e.g. the RCA C70127 and C70128) and commonly have sensitivities in the near ultraviolet several orders of magnitude lower than in the vacuum ultraviolet. The selective photocathodes in these tubes are either metallic (nickel, tungsten, gold, etc.) or composite surfaces of materials such as the alkali tellurides and halides or copper iodide. In general in the range 2000–1000 Å the metallic photocathodes have much lower quantum efficiencies, although tungsten may yield a quantum efficiency of up to 10% at the shorter wavelength end.[26] Most of the composite cathode tubes are of the sealed off type to prevent contamination or decomposition of the photosurface. Thus their use is limited to about 1100 Å. At shorter wavelengths it becomes necessary to use a windowless tube and a photocathode that will stand repeated exposure to air. Fortunately, for several metals (e.g. Cd, W) a strong volume photoelectric effect sets in below about 1300 Å, and yields high quantum efficiencies in the 1300–500 Å region.[26, 27] Thus the combination of composite windowed tube and pure metallic windowless tube gives good coverage of the whole vacuum ultraviolet region with adequate sensitivity and good stray light rejection properties.

TABLE 3.2
UV and IR window materials

	Useful transmission (microns)		Crystal structure	Melting point (°C)	Sol. in H_2O at RT (g/100 g H_2O)
	$\lambda_{min.}$	$\lambda_{max.}$			
LiF	0·105	7	cubic-NaCl	848	0·27
MgF_2	0·11	7·5	tetragonal-rutile	1255	0·01
CaF_2	0·125	9	cubic	1418	0·16
BaF_2	0·14	15	cubic-CaF_2	1320	0·001
SrF_2	0·12	10	cubic-CaF_2	1190	0·01
Al_2O_3	0·14	6	hexagonal	2050	0
SiO_2	0·15	5	hexagonal	<1470	0
NaCl	0·18	16	cubic	801	36
KCl	0·19	20	cubic-NaCl	776	35
KBr	0·20	25	cubic-NaCl	730	54
KRS5	0·5	40	cubic	414	0·05
CsBr	0·2	40	simple cubic	636	124
CsI	0·2	50	simple cubic	621	44
NaF	0·17	11	cubic-NaCl	997	4·2
AgCl	0·4	25	cubic	455	0
Si	1·1	20	cubic-diamond	1420	0
Ge	1·7	20	cubic-diamond	959	0
InSb	7·0	18	cubic-zincblende	523	0
Pyrex	0·3	4	–	–	0

3.2.4. *Windows and filters*

In this section we have included a table of the more common materials used over the infrared to vacuum ultraviolet region. Table 3.2 shows the range of transparency, crystal structure, melting-point and the solubility in water which is a measure of their resistance to attack by atmospheric water vapour. Some of the more common semiconductors have been included, the short wavelength cut off in all cases being the fundamental absorption edge of the material. The long wavelength limits are approximate values for undoped material and may vary in practice.

To the list in Table 3.2 must be added the very extensive range of special filters covering the ultraviolet, visible and near infrared regions which are available commercially. Pigmented glass filters of all types (i.e. low pass, high pass, band pass and band rejection) can be obtained from manufacturers such as Corning in the U.S.A. and Schott in Europe; such filters tend to possess gradual cut-offs. Multilayer interference filters with their characteristic extremely sharp cut-offs are also readily available, and many manufacturers besides producing a range of these, are prepared to tailor-make interference filters to meet particular specifications.

Grey filters, used for attenuators, are available commercially either in a pigmented glass form, or as a thin metallic film. Such filters can be readily made in the laboratory by evaporating a thin aluminium layer on to a quartz, glass or other suitable substrate.

3.2.5. *Mirrors*

Over the larger part of the spectral region we consider here, aluminized mirrors are clearly the obvious choice. Freshly evaporated aluminium possesses an extremely high reflectance at normal incidence from the far infrared to the vacuum ultraviolet. The reflectance is not constant, however, and there is a minimum in the near infrared. Surface oxidation has a pronounced deleterious effect on the reflectance below 2200 Å. This effect is illustrated in Fig. 3.2 which shows the large effect of ageing at short wavelengths. Unprotected aluminium surfaces are thus of limited practical use below about 1500 Å, and in addition to this, long-term stability is not found below about 3000 Å. Overcoating with a thin protective layer (e.g. SiO_2, MgF_2 or

LiF) both improves the stability over long periods and enhances the short wavelength reflectivity. Overcoating must be done immediately after aluminizing and without breaking the vacuum. A detailed description of both the preparation and properties of LiF and MgF_2 coated mirrors for the vacuum ultraviolet has been given by Hass and Hunter.[28] With overcoating, one can obtain rather stable mirrors possessing normal incidence reflectivities of about 80% at 1200 Å; this now makes it possible to consider using more complex reflecting optics in vacuum ultraviolet systems, at least down to the LiF cut-off at 1050 Å.

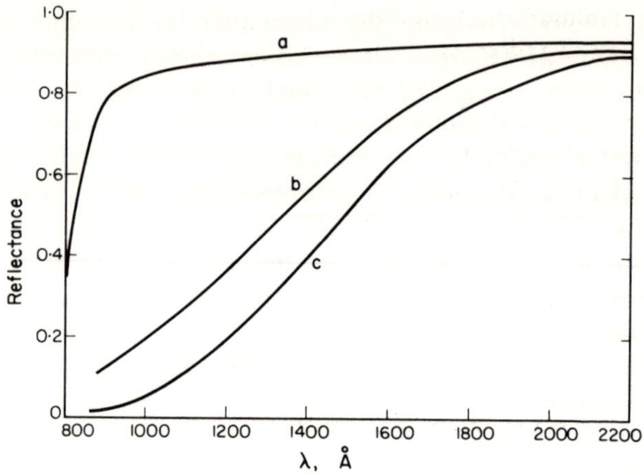

FIG. 3.2. Normal incidence reflectance of Al under different conditions: (a) Freshly deposited Al, measured *in vacuo* without exposure to air. (b) Evaporated Al after 24-hr exposure to air. (c) Evaporated Al after several months' exposure to air (Hass, Hunter and Tousey[13]).

Below this wavelength reflectances are much lower: ZnS reflects 15% at 800 Å and Pt, which is probably the best reflecting surface for use below about 1000 Å, has a reflectivity of about 20% at 600 Å, falling to a value of a few per cent at 300 Å. Clearly, below these wavelengths, normal incidence systems become unusable, and it becomes essential to work at grazing incidence.

3.2.6. *Polarizers*

Infrared polarizers are mainly of the Brewster angle type consisting of a stack of several transparent plates or thin films at the correct angle. Such polarizers can conveniently be made from germanium or silicon where the high refractive index allows them to be used in reflection as well as transmission. An efficient transmission polarizer using thin unsupported evaporated selenium films has been described by Elliott, Ambrose and Temple.[29] Probably the most widely used infrared polarizer is that made from a pile of about six rolled 0·1–0·5 mm thick AgCl plates; the polarization is better than 98% and the plates can easily be cold-rolled from thicker material. AgCl polarizers must be kept in darkness when not in use, as prolonged exposure to visible and ultraviolet light causes discoloration. AgCl is also attacked by common metals very rapidly, and the polarizing plates must be mounted in nonmetallic holders. The beam deviation inherent in Brewster angle polarizers can be eliminated by constructing the polarizer from two stacks of plates each at the Brewster angle, but set in opposing directions. A recently developed infrared polarizer consists of oriented needle-shaped inclusions of NiSb dispersed regularly in a matrix of InSb; this polarizer can be used from $\lambda = 8\ \mu$ to about 50 μ and is produced by Siemens, Germany.

Polaroid sheet provides a cheap and reasonably efficient polarizer in the visible region. Its usable range can be extended somewhat further into the ultraviolet by removing the protective gelatine layers with acetone. The most efficient visible polarizers are those utilizing doubly refracting crystals such as calcite and quartz (e.g. Nicol, Rochon, Glan–Thompson and Wollaston prisms), where the unwanted ray is removed either by internal reflection or by masking. Care must be taken in using these polarizers, as the emergent ray is usually displaced with respect to the incident ray (the Rochon prism is an exception to this). The use of air gap prisms instead of the usual Canada balsam cemented polarizers can considerably extend the range into the ultraviolet. The air gap calcite Glan prism is usable down to 2100 Å.

LiF is the most convenient polarizing material for the vacuum ultraviolet. The Harshaw Company now manufactures LiF with a guaranteed transmission at 1200 Å and thus it is now feasible to construct efficient Brewster angle stacks which can be used down to about 1100 Å. Several workers have obtained good results using a single LiF Brewster angle reflection despite the low reflected intensity. A polarizer of this type, incorporated into an attachment for reflectivity measurement in the vacuum ultraviolet, is described in the following section. An alternative vacuum ultraviolet polarizer is an air gap Glan prism made from MgF_2; such a polarizer should be usable down to about 1200 Å. Recently, a naturally occurring iron rich silicate – biotite – which has micaceous appearance and cleavage properties, has been demonstrated to show efficient polarization by Brewster angle reflection over the range 1100–6000 Å.(30) This type of polarizer appears to be more efficient than LiF, and its polarization effects may well extend below 1100 Å.

Below the LiF cut-off, apart from the electron synchrotron, no efficient wide band polarizer has been demonstrated at present, although some suggestions have been made for obtaining partial polarization by reflection from metallic surfaces.(31)

3.3. Measurement techniques

3.3.1. *Reflectance under oblique incidence*

For the determination of the optical constants by this method several choices of two independent measurements may be made. As we pointed out in Section 2.2.2, R_p and R_n can be determined at one angle of incidence, the ratio R_p/R_n can be measured at two angles of incidence (thus avoiding the measurement of absolute intensities), or the reflectance at two angles can be determined for incident light having a known state of polarization. For this latter method, light polarized in the plane of incidence gives the greater sensitivity. In all these experiments it is desirable to make the measurement near the Brewster angle where R_p is most sensitive to change in angle. Figure 3.3 shows schematically how such measurements can be made. Provision

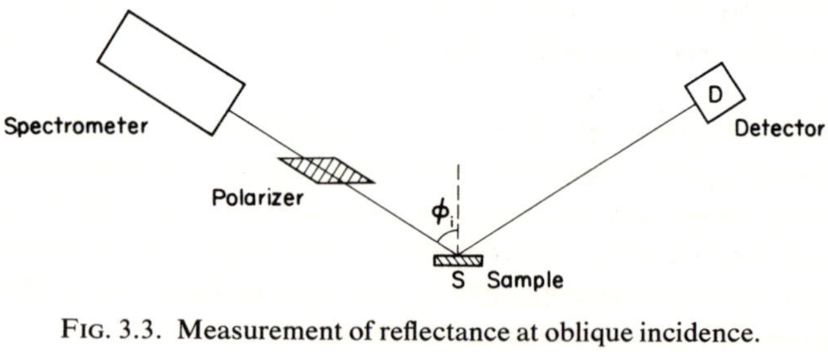

FIG. 3.3. Measurement of reflectance at oblique incidence.

should be made for both moving the sample out of the beam and rotating the sample and detector about an axis through the sample, so that absolute reflected intensities can be measured and the angle of incidence can be varied. A simple gearing mechanism to give a 2:1 ratio between detector and sample rotations is a convenient addition to this device.

3.3.2. *Measurement of azimuth and phase angle*

Basically this method requires only a polarizer in the incident beam, as in Fig. 3.3, and an analyser in the reflected beam. The reflected intensity after passing the analyser depends on the azimuth angles of the polarizer and analyser relative to the plane of incidence; intensity measurements at several settings of polarizer and analyser are sufficient to determine both the azimuth and phase angle of the sample. Details of this and several other ellipsometric methods are described by Fan.[32]

3.3.3. *Measurement of normal incidence reflectivity*

Two basic methods are available for the measurement of normal incidence reflectance; they may be termed the relative and absolute methods. The expression "normal incidence" applies here to measurements taken with incidence angles less than 15° to the specimen normal. In practice the error introduced using this criterion is very small (Fig. 2.1). In the relative method, reflectivity is measured by comparison with a reference mirror, usually aluminium, the reflectivity of which must also be known if absolute results are required. In general relative measurements are mechanically simpler to perform. In the infrared, where aluminium has a high and constant reflectivity, a relative method using a reference mirror is quite satisfactory (it may even be preferred for some purposes), but for visible and ultraviolet wavelengths it is essential to be either able to take absolute reflectivity measurements directly, or to have a facility for checking the calibration of the reference mirror at frequent intervals. Aluminium mirrors are available commercially which have a protective evaporated coating of SiO_2 to provide durability (e.g. the Alflex series of mirrors made by Balzers of Liechtenstein); these mirrors are extremely stable, and can be used down to 2000 Å, but nevertheless each mirror must be calibrated separately for use below about 2500 Å.

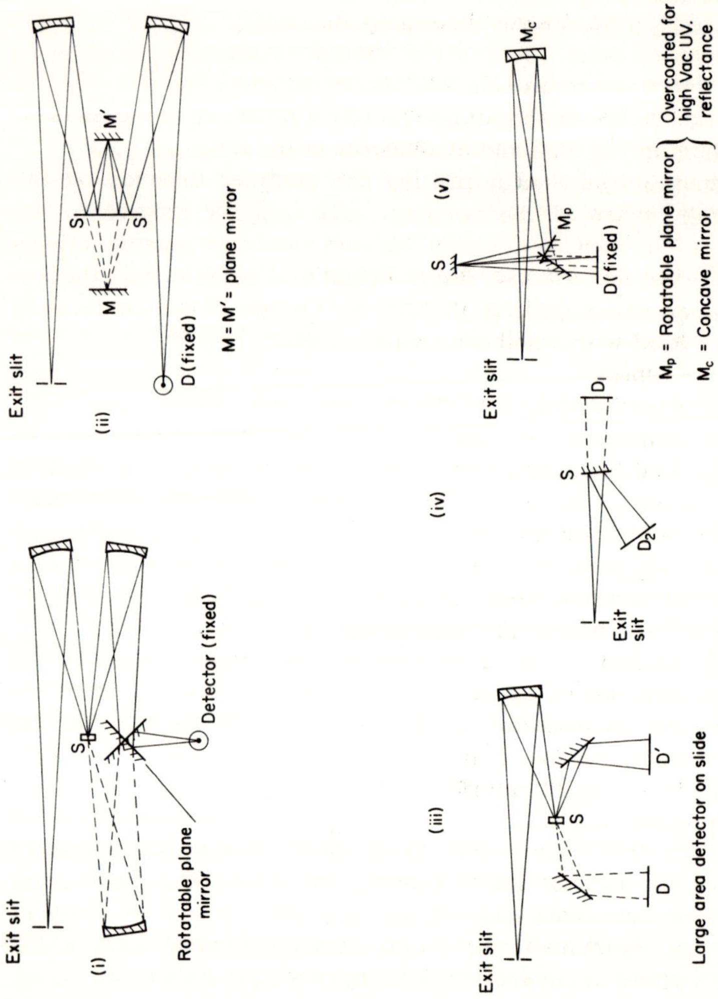

FIG. 3.4. Several different methods for the measurement of normal incidence reflectance.

There exist a number of different ways of taking absolute reflectivity measurements, the basic method being illustrated by Fig. 3.3 where the detector can be rotated about the sample to record the incident and reflected intensities. Clearly for measurements on cubic materials the polarizer is unnecessary as both R_p and R_n tend towards the same values as the angle of incidence approaches zero. Some of the alternative methods for absolute reflectance determination are shown schematically in Fig 3.4. Most of these are also suitable for absorption measurements; all can be used for relative reflectance measurements if the sample is replaceable by a plane mirror. Specific points about these systems are given below.

(i) Suitable for use in the infrared where the existence of several reflecting surfaces is possible. This system can be used with small area infrared detectors such as thermocouples.

(ii) Also suitable for infrared use, but needs rather large samples or two smaller samples. It is particularly useful for determining the reflectance of high-reflectivity materials such as metals, as the measurement gives values of R^2 rather than R.

(iii) This system is intended primarily for use with existing commercial single-beam spectrophotometers such as the Zeiss or Beckman instruments, where two interchangeable detectors (e.g. a photomultiplier and a photocell) are mounted together in a somewhat bulky unit which is detachable from the main spectrometer. Such a unit can conveniently be mounted on a slide moving between the reference and sample positions.

(iv) This simple twin detector arrangement is intended for use in the vacuum ultraviolet where mirror reflections are undesirable. The detectors, which of course must be an integral part of a vacuum housing mounted on the spectrometer exit slit, are commonly phosphor-coated 1P21 photo-multipliers.

(v) This system is intended for use in the vacuum ultraviolet down to a wavelength of about 1000 Å. It employs LiF or MgF_2 overcoated mirrors and besides possessing a beam focus which allows the use of small specimens has the advantage that both sample and detector remain fixed during measurement. The reference and sample intensities are obtained by simply rotating the small plane mirror between fixed stops. The system can in addition be used for measurements with plane polarized radiation if the plane mirror is replaced by a LiF plate adjusted to reflect at the Brewster angle for both sample and reference positions.

The measurement of low-temperature reflectivity presents a problem at all wavelengths. Surface contamination with ice, grease, diffusion pump oil or any other condensable vapour provides a source of error causing progressively greater difficulties as the wavelength becomes shorter. A secondary problem arises in the infrared as a consequence of the hygroscopic nature of cryostat window materials such as KBr and KCl: after a period depending on the ambient humidity, the windows in a cryostat may have different transmission characteristics. When working with such windows it is better to make relative measurements and mount a calibrated aluminium reference mirror beside the sample in a cryostat where the sample holder can be moved while the windows remain stationary. (The reflectivity of aluminium does not change appreciably with temperature in the infrared and remains uniformly high at about 98%.) This problem does not arise in the visible and near ultraviolet regions where quartz provides a convenient stable window material.

Surface contamination at 77°K by condensable vapours can be avoided completely over the whole wavelength region down to about 1800 Å by surrounding the sample with a shield in thermal contact with the sample holder. This shield, with the appropriate windows, should be a good mechanical fit over the sample holder, but should not be vacuum-tight. Contaminants will tend to deposit on the outside of the shield windows where their effect if at all significant will cancel out. Both the shield windows and the exterior windows of the cryostat should be inclined at a sufficiently large angle to the specimen surface so that direct reflections from these windows do not enter the detector. In the range 1800–1050 Å contamination of the sample surface at 77°K can be minimized but not avoided completely using shields of this type, provided the vacuum in the cryostat is around 10^{-6} mm or better.

Under these conditions, errors of at least a few per cent over a period of an hour can be expected. Scouler[33] has described a vacuum ultraviolet cryostat employing LiF windows for measurements at 4°K. The construction is more complex than the simple shield arrangement detailed above, but gives excellent stability and freedom from contamination. The use of windows, of course, limits the range to 1050 Å; below this wavelength no low-temperature reflectivity measurements on semiconductors or insulators have been reported, but such measurements should be feasible if a sufficiently high vacuum can be maintained in the sample chamber.

3.3.4. *Measurement of normal incidence absorption and reflectance*

Most of the methods described above are also applicable to the simultaneous measurement of absorption and reflectance. For semiconductors in the region of the fundamental absorption edge, while the reflectance is a slowly varying function of wavelength, the absorption coefficient varies over several orders of magnitude. Calibrated grey filters allow a range of about 10^5 in transmission to be measured on one sample if the sample thickness is chosen correctly. In such a measurement it is essential to use a suitable short-wavelength pass filter to reduce the effect of scattered light in the region where the material is transparent.

Absorption measurements at low temperature present no serious problem provided a reasonable vacuum (10^{-5} torr) is maintained in the cryostat. The criteria for cleanliness in absorption measurements are far less stringent than those for reflectance. In the visible region measurements can be made with the sample completely immersed in the coolant to obtain thermal equilibrium but this method can introduce serious errors if reflectance is also being measured. If only transmission is being measured, and surface contamination is suspected (e.g. in the case of measurements on evaporated films), a differential measurement using two samples of different thickness can prove useful.

3.3.5. *Measurement of pressure and strain effects*

Determination of optical properties under hydrostatic pressure must be made in a pressure cell equipped with a suitable window. At high pressures (~ 5000 atm) sapphire has proved to be a convenient material. The hydrostatic fluid must be stable over the range of the measurement; nitrogen gas, silicone oils and Freon have all been used, although Freon solidifies at high pressure and can only be used to ~ 1500 atm. The experimental techniques involved in hydrostatic pressure measurements have been described by Neuringer[34] and Michels *et al.*[35]

Uniaxial stress can be applied by applying a bending moment to filamentary shaped samples. Specimens can either be held in jaws at both ends[36] or mounted in a four-point bending jig to ensure uniform application of stress.[37] For piezoreflectance experiments these methods have the advantage that both longitudinal extension and compression can be observed on the same sample by viewing the two sides. In measurements of this type (also in hydrostatic pressure experiments) it is important to take readings with both increasing and decreasing strain. Large hysteresis effects imply the presence of voids and imperfections within the material. Two highly sensitive methods of investigating piezoreflectance of metals and semiconductors using transducers to introduce dynamic strain into the material have been recently described by Engeler *et al.*[38] and Gobeli and Kane.[39]

3.3.6. *Measurement of electroreflectance*

The electroreflectance measurements on semiconductors and insulators reported in Chapter 9 involve the investigation of the normal incidence reflectivity in the fundamental absorption region when an electric field is applied perpendicular to the reflecting surface. Seraphin and Hess[40] have employed a condenser-like system for such measurements. A thin (0·01 mm) Mylar dielectric is sandwiched between a transparent conductive coating on quartz, which serves as a field electrode, and the semiconductor surface. The sandwich is cemented together with a lens cement which matches the refractive indices of the transparent components as well as possible. A small d.c. bias is applied to the system to enable the surface potential to be varied and an a.c. field (250 Hz and up to 3×10^5 V/cm) is used to modulate the reflectance. A narrow band amplifier and oscilloscope are used to display the 250 Hz modulation impressed onto a monochromatic beam reflected at the semiconductor surface.

An alternative method is to apply the alternating field at a semiconductor–electrolyte interface. Shaklee, Pollak and

Cardona[41] have used dilute aqueous KCl as an electrolyte and obtained reflectance modulations as high as 0·2% for an applied a.c. voltage of 2 V peak to peak. Again, the samples should be biased with a d.c. voltage to form a blocking contact at the surface. The electrolyte technique permits irregularly shaped samples to be measured (e.g. cleaved material) and also permits effects such as uniaxial stress to be investigated.

3.3.7. *Measurement of photoemission*

Photoemission occurs when the absorption of a photon leads to ejection of an electron from an absorbing material into the vacuum. For semiconductors and insulators either the photoelectric yield may be measured as a function of the wavelength of the incident light, or the energy distribution of the emitted electrons may be measured at a fixed wavelength. Both experiments involve the preparation of ultra-clean surfaces. In order to see as much of the interband optical structure as possible in the photoelectric yield it is desirable to lower the vacuum level below that of a clean surface and this may be accomplished by depositing a thin layer of caesium on the surface to be studied. Details of the application of this technique to silicon are given by Spicer and Simon;[42] studies on CdS (without deposition of caesium) have been made by Spicer and Kindig.[43] For a detailed discussion of the experimental techniques in photoemission studies the reader is referred to the review by Sommer and Spicer.[44]

3.4. Preparation of samples

Surface preparation clearly plays a critical role in the measurement of a parameter such as the reflectivity. The difficulties in obtaining reproducible surfaces which reveal the properties of the bulk material become more severe at shorter wavelengths. The absorption coefficient of semiconductors in the fundamental region rises to values of the order of 10^6 cm^{-1}, thus the penetration depth of the radiation is at most a few hundred angstroms; the surface damage that can be tolerated must extend to a depth considerably less than this. Also the microscopic surface roughness must be less than the wavelength of the radiation being used, otherwise scattering effects will be introduced. Precautions must be taken to prevent the formation of excessive oxide films, as these can cause totally misleading interference effects quite apart from any intrinsic absorption of the oxide itself. For reflectivity measurements the most satisfactory solution is to measure a freshly cleaved single crystal face and fortunately this is possible for most of the common semiconductors and insulators. It is not necessary for the cleaved face to be optically flat, although if this can be achieved it clearly simplifies the measurement. Provided that accurate values are available for the reflectivity at one wavelength (for example, from refractive index data in a transparent region) the rest of the curve can be fitted to this value.

When cleavage is neither possible nor convenient as in the case of measurements on anisotropic materials on crystal faces other than easy cleavage faces, measurements on polycrystalline materials, or preparation of samples for an absorption measurement, the surfaces must be polished mechanically and possibly also etched. While the etching step can be omitted when preparing samples for most absorption measurements, it is essential for a reflectivity measurement. Mechanical polishing whether by hand or machine should be performed in several stages using progressively finer abrasives. For anisotropic materials such as Bi_2Te_3 or CdI_2 which cleave extremely easily, and where one requires a polished surface perpendicular to the cleavage planes, it is advisable to first embed the sample in an epoxy resin. For absorption measurements, the minimum thickness obtainable by mechanical polishing depends on the material, but for germanium and silicon, polished slices down to about 5 μ thick over an area of 1 cm^2 or more are possible without undue difficulty if the slice is waxed on to a holding plate after the first side has been polished. Etching the mechanically polished sample ideally removes the surface damage introduced by the polishing procedures. An extremely wide choice of chemical and electrolytic etches is available, and for further details the reader is referred to two useful reference books on the subject.[(45)]

A different approach to sample preparation, particularly when extremely thin specimens are required, as in fundamental region absorption measurements on semiconductors, is to evaporate the material epitaxially onto a single crystal substrate. The method is capable of extremely good results if a transparent substrate is available with the same symmetry and approximately the same lattice parameter (±5%) as the material to be evaporated. In some cases, however, good crystalline films may be obtained if these conditions are not fulfilled.[46] This method can be applied to many of the common semiconductors, but is particularly useful when measurements are required over an alloy system of two components, as the samples can often be prepared in a simple evaporation system without recourse to elaborate crystal-growing equipment. In practice the substances to be evaporated can either be placed in separate resistance heated boats or mixed in the required proportions in a single boat. The substrate should be freshly cleaved single-crystal material, and should be held at a temperature which has been found empirically to be between one-half and two-thirds of the Centigrade melting point of the substance to be evaporated. This temperature is necessary to give the molecules reaching the substrate sufficient mobility to find an appropriate lattice site, and yet still retain a finite sticking coefficient. The technique has been applied successfully to various alloys of the alkali halides[47] and in the case of single compounds to the lead salts PbS, PbSe and PbTe,[10] and to Bi_2Te_3.[48]

CHAPTER 4

INTERBAND TRANSITIONS AND CLASSIFICATION OF ELECTRONIC STATES

4.1. Quantum theory of interband transitions

In Chapter 2 we have given the phenomenological description of the optical properties in terms of the complex frequency-dependent dielectric constant $\epsilon(\omega)$. We now proceed to derive the relation between $\epsilon(\omega)$ and the electronic excitation in the solid by the electromagnetic wave. The discussion will be restricted to interband transitions in the one-electron approximation, i.e. to transitions between states which are not in the same band.

The transition probability for an electron from an occupied valence band state to an empty conduction band state results from the perturbation of the system by the time-dependent vector potential of the electromagnetic field. The total Hamiltonian for an electron is then

$$H = \frac{1}{2m}(\mathbf{p} + e\mathbf{A})^2 + V_c(\mathbf{r}),$$

where $\mathbf{p}$ is the momentum, $V_c(\mathbf{r})$ the periodic crystal potential and $\mathbf{A}$ the vector potential. For a plane wave described by equations (2.1) and (2.2) we write

$$\mathbf{A} = A_0 \mathbf{e}\, e^{i(\boldsymbol{\kappa}\cdot \mathbf{r} - \omega t)} + \text{c.c.} \tag{4.1}$$

where c.c. denotes the complex conjugate term and $\mathbf{e}$ is the unit vector of polarization in the direction of the electric field. The first-order perturbation operator is then the linear term

$$H'(\mathbf{r},t) = \frac{e}{m}\mathbf{A}\cdot\mathbf{p}.$$

For the absorption we only have to consider the first term in equation (4.1), the complex conjugate term describes the emission. The transition probability for an electron from a valence band state with wave vector **k** to a conduction band state with wave vector **k**′ is given by[(49)]

$$w(\omega,t,\mathbf{k},\mathbf{k}') = \frac{e^2}{m^2\hbar^2}\left| \int_0^t \mathrm{d}t' \int_V \mathrm{d}\mathbf{r}\psi_c^*(\mathbf{k}',\mathbf{r})\mathbf{A}\cdot\mathbf{p}\psi_v(\mathbf{k},\mathbf{r})\right|^2. \quad (4.2)$$

Since ψ_v is a Bloch type eigenfunction belonging to E_v we can write

$$\psi_v = \mathrm{e}^{-i\frac{E_v(\mathbf{k})}{\hbar}t}\,\mathrm{e}^{i\mathbf{k}\cdot r}u_v(\mathbf{k},\mathbf{r}), \quad (4.3)$$

where the part u_v has the periodicity of the lattice. An equivalent expression holds for ψ_c. These two equations and equation (4.1) are put into (4.2) and give

$$w(\omega,t,\mathbf{k},\mathbf{k}') = -\left(\frac{e}{m}\right)^2 A_0^2\left| \int_0^t \mathrm{d}t'\mathrm{e}^{\,i\frac{E_c(\mathbf{k}')-E_v(\mathbf{k})-\hbar\omega}{\hbar}t'} M_{vc}\right|^2, \quad (4.4)$$

where M_{vc} is an integral over all space

$$M_{vc} = \int_V \mathrm{d}\mathbf{r}\;\mathrm{e}^{-i(\mathbf{k}'-\boldsymbol{\kappa})\cdot\mathbf{r}}\; u_c^*(\mathbf{k}',\mathbf{r})\mathbf{e}\cdot\nabla\mathrm{e}^{i\mathbf{k}\cdot r}u_v(\mathbf{k},\mathbf{r}). \quad (4.5)$$

In time-dependent perturbation theory this leads to the transition probability per unit time and unit volume[(49)]

$$W_{vc}(\omega) = -4\pi\hbar\left(\frac{e}{m}\right)^2 A_0^2 \int \frac{\mathrm{d}\mathbf{k}}{4\pi^3}\left|M_{vc}\right|^2 \delta(E_c - E_v - \hbar\omega). \quad (4.6)$$

The δ-function in equation (4.6) represents the first selection rule based on energy conservation. The transition probability is different from zero only if $E_c - E_v = \hbar\omega$, i.e. if the energy difference between the empty state E_c and the occupied state E_v is equal to the energy of the photon. The second selection rule based on conservation of momentum follows from equation (4.5). The matrix element M_{vc} will vanish unless $\mathbf{k}' = \mathbf{k} + \boldsymbol{\kappa}$. This is a consequence of the translational invariance of u_c and u_v and can easily be verified by putting $\mathbf{r}+\mathbf{R}$ instead of $\mathbf{r}$, where $\mathbf{R}$ is a lattice vector. The wave vector of the radiation $\boldsymbol{\kappa}$ is very small compared with Brillouin zone dimensions and can be neglected. Thus only vertical transitions are allowed in this dipole approximation and we can take $\mathbf{k}=\mathbf{k}'$.

The energy loss per unit volume of the plane wave by absorption is $W_{vc}\hbar\omega$. It can also be described by $\frac{1}{2}\sigma E_0^2$, where $\mathbf{E}=-\partial\mathbf{A}/\partial t$. With equation (2.6) we have

$$\begin{aligned}\epsilon_2 &= \frac{\sigma}{\omega\epsilon_0} = \frac{2\hbar W_{vc}}{\epsilon_0 E_0^2} = \frac{2\hbar W_{vc}}{\epsilon_0\omega^2 A_0^2} \\ &= \frac{8\hbar^2\pi e^2}{\epsilon_0 m^2\omega^2}\int\frac{\mathrm{d}\mathbf{k}}{4\pi^3}\left|M_{vc}\right|^2\delta(E_c-E_v-\hbar\omega).\end{aligned} \tag{4.7}$$

This is a very important result. It shows that the imaginary part of the dielectric constant at a frequency ω is given by all vertical interband transitions between states of energy difference $\hbar\omega$ and by the matrix element M_{vc} which depends on the wave functions of initial and final states. If both these quantities are known, i.e. if all the electron eigenstates and their wave functions are known, ϵ_2 can be calculated. As outlined in Section 2.3, ϵ_1 can then be obtained from ϵ_2 by use of the dispersion relations and we have a relation between the dielectric constant or the optical constants and the electronic band structure of a material.

Because of the dipole approximation applied in this derivation we obtain only the frequency dependence of ϵ. A more general approach has been taken by Ehrenreich and Cohen[50] with the self-consistent field method. They calculate the wave-number and frequency-dependent dielectric function $\epsilon(\mathbf{q},\omega)$ describing the response of the medium to an electromagnetic perturbation $E(\mathbf{r},t)$. The field $E(\mathbf{r},t)$ can be of any space and time dependence and is not restricted to the sinusoidal form assumed in equation (4.1). In the limit $\mathbf{q}\to 0$ the results of Ehrenreich and Cohen for the interband contribution to $\epsilon(\mathbf{q},\omega)$ reduce to the frequency dependent dielectric function

$$\epsilon(\omega)=1-\frac{e^2}{m\pi^2}\int \mathrm{d}^3\mathbf{k}\sum_{ll'} f_l(\mathbf{k})f_{l'l}\left(\omega-\omega_{ll'}+\frac{i}{\tau_{ll'}}\right)^{-1}\left(\omega+\omega_{ll'}+\frac{i}{\tau_{ll'}}\right)^{-1}. \tag{4.8}$$

$f_l(\mathbf{k})$ is the Fermi distribution, and $f_{l'l}$ is the oscillator strength for the transition between bands with indices l and l' which is proportional to the square of the matrix element defined in equation (4.5). The frequency $\omega_{ll'}$ is defined by $\hbar\omega_{ll'} = E_l(\mathbf{k}) - E_{l'}(\mathbf{k})$ where $E_l(\mathbf{k})$ is the energy of an electron having wave number $\mathbf{k}$ in band l. Lifetime broadening effects have been introduced by incorporating the complex frequency $(\omega + i/\tau_{ll'})$ into equation (4.8). The relaxation time $\tau_{ll'}$ depends on the bands involved in the transition but in the present situation it is assumed that $\omega\tau_{ll'} > 1$. The imaginary part $\epsilon_2(\omega)$ of $\epsilon(\omega)$ in equation (4.8) is equivalent to the formula (4.7).

4.2. Density of states

If M_{vc} is assumed to be slowly varying with $\mathbf{k}$, the integral in equation (4.7) can be rewritten. Using the property of the δ-function:

$$\int_a^b g(x)\delta[f(x)]\,\mathrm{d}x = \sum_{x_0} g(x_0)\left|\frac{\partial f}{\partial x}\right|^{-1}_{x=x_0}$$

where $f(x_0) = 0$ and $a < x_0 < b$. In the three-dimensional case one obtains

$$\epsilon_2 = \frac{8\hbar^2\pi e^2}{\epsilon_0 m\omega^2}\int_S \frac{1}{4\pi^3}\frac{\mathrm{d}S|M_{vc}|^2}{|\nabla_\mathbf{k}(E_c - E_v)|_{E_c - E_v = \hbar\omega}},$$

where $\mathrm{d}S$ is a surface element on the equal energy surface in $\mathbf{k}$-space $E_c(\mathbf{k}) - E_v(\mathbf{k}) = \hbar\omega$. Detailed calculations in many cases have shown that M_{vc} is indeed a slowly varying function compared with

$$J_{vc} = \int_S \frac{1}{4\pi^3}\frac{\mathrm{d}S}{|\nabla_\mathbf{k}\,(E_c - E_v)|_{E_c - E_v = \hbar\omega}} \tag{4.9}$$

the so-called joint density of states. J_{vc} is a strongly varying function of $\mathbf{k}$ where $\nabla_\mathbf{k}\,(E_c - E_v) = 0$. The energy difference function $(E_c - E_v) = E_{vc}(\mathbf{k})$ in a periodic crystal is a periodic function in the reciprocal lattice and must therefore also possess such points, termed critical points, just as a consequence of translational invariance and symmetry. Critical point behaviour and its important consequences for the phonon density of states in a crystal were discussed first by van Hove[51] and Phillips[52]. Similar considerations hold for the electronic states and it is necessary to distinguish between two types of critical points corresponding to the different conditions

$$\nabla_{\mathbf{k}} E_c(\mathbf{k}) = \nabla_{\mathbf{k}} E_v(\mathbf{k}) = 0 \tag{4.10}$$

and

$$\nabla_{\mathbf{k}} E_c(\mathbf{k}) = \nabla_{\mathbf{k}} E_v(\mathbf{k}) \neq 0. \tag{4.11}$$

Critical points of the first type occur only at highly symmetrical points of the Brillouin zone and their existence is predictable from symmetry alone. Critical points of the second type can exist at points of lower symmetry in **k**-space at which the symmetry does not require the existence of critical points of the first type and cannot be predicted solely from the symmetry. The analytic behaviour of $J_{vc}(\omega)$ near critical points can best be seen by expanding $(E_c - E_v)$ in a Taylor series about the critical point $\mathbf{k}_0$ of energy difference E_0

$$E_c - E_v = E_0 + \sum_{i=1}^{3} a_i (k_i - k_{0i})^2 \tag{4.12}$$

and performing the integral in equation (4.9). According to the sign of the a_i's we obtain four kinds of critical points M_0 to M_3 which are listed in Table 4.1 together with the energy dependence of J_{vc}. The different constants C_i or $C_{i'}$ depend on the absolute values of the a_i's.

TABLE 4.1
Joint density of states for different critical points

Type of c.p.	Notation	a_1	a_2	a_3	J_{vc}	
					$E < E_0$	$E > E_0$
Minimum	M_0	+	+	+	0	$C_0(E-E_0)^{1/2}$
Saddle point	M_1	+	+	−	$C_1 - C_1'(E_0-E)^{1/2}$	C_1
Saddle point	M_2	+	−	−	C_2	$C_2 - C_2'(E-E_0)^{1/2}$
Maximum	M_3	−	−	−	$C_3(E_0-E)^{1/2}$	0

M_0 and M_3 (equation (4.10)) give the well-known square-root dependence of the density of states. M_1 and M_2 (equation (4.11)) will be shown to be the dominant critical points in the interband spectra. The total numbers of the four different kinds of critical points are related by the Morse relations.[52]

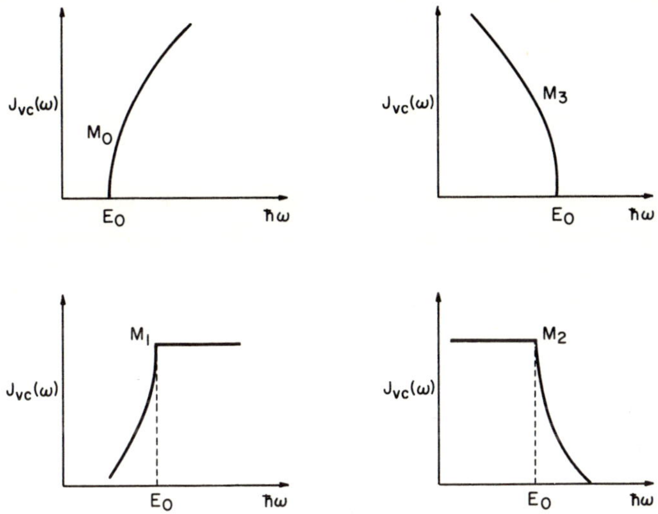

FIG. 4.1. Joint density of states plotted against energy for different critical points in three-dimensional crystals.

The expected behaviour of J_{vc} as a function of energy is qualitatively shown in Fig. 4.1. We now see that these characteristic line shapes give us a valuable method for identifying the nature of the interband transitions which are experimentally observed in real materials. The behaviour of J_{vc} shown in Fig. 4.1 illustrates another important point: a single critical point does not in itself produce a narrow peak in the density of states and thus in ϵ_2, but only an edge. To obtain a peak due solely to interband transitions, there must be two singularities M_1 and M_2 accidentally having nearly the same critical point energy E_0.

In the case of two-dimensional crystals the **k**-vector in reciprocal space has two components so that the critical points depend only on two coefficients a_1 and a_2 in the Taylor expansion of equation (4.12). We have the minimum and maximum type as in the three-dimensional case but only one kind of saddle point from

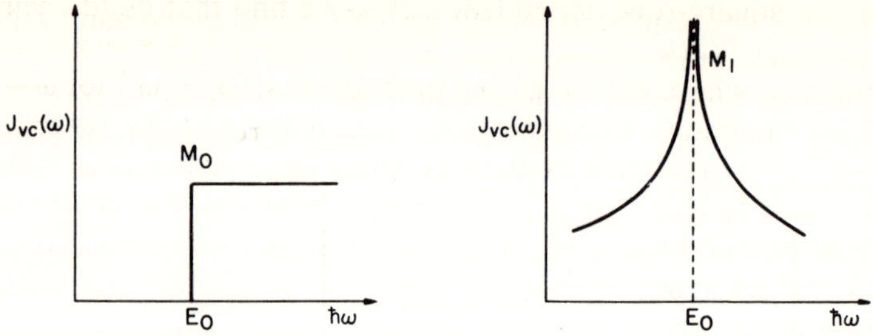

FIG. 4.2. Joint density of states plotted against energy for M_0 and M_1 critical points in two-dimensional crystals (Bassani[49]).

opposite signs of a_1 and a_2. The J_{vc} behaviour is, however, very different from the functions in Fig. 4.1. For the extremal types given by equation (4.10) J_{vc} is just a step function at E_0. For the saddle point J_{vc} takes the form:[49]

$$J_{vc} = C_1\{\ln[C_2 + (|E_0 - \hbar\omega| + C_2^2)^{1/2}] - \tfrac{1}{2}\ln|\hbar\omega - E_0|\} \quad (4.13)$$

which gives a logarithmic singularity at $\hbar\omega = E_0$. The expected behaviour of J_{vc} is shown in Fig. 4.2. This two-dimensional model assumes that the bands involved have equal masses in both the x- and y-directions. The joint density of states for the two-dimensional case may be an appropriate description of electronic states in layer compounds or of surface states.

4.3. Behaviour of the optical constants near interband edges

In Section 4.2 we have seen that the joint density of states near an M_0 type critical point varies as $(E-E_0)^{1/2}$ for $E > E_0$. Over a small energy interval above E_0 the product $\omega^{-2}|M_{vc}|^2$ can be regarded as slowly varying compared with J_{vc} so that ϵ_2 and also K have the same energy dependence as J_{vc}. Since $J_{vc} = 0$ when $E < E_0$ this represents an absorption threshold at E_0 followed by a strong increase in ϵ_2. From the structural correlations implied in the dispersion relations discussed in Section 2.3 we know that ϵ_1 has a peak where ϵ_2 increases strongly. Explicitly this can be seen if equation (2.26) is integrated by parts and expressed in the form:

$$\epsilon_1(\omega) = 1 + \frac{1}{\pi}\int_0^\infty \frac{d\epsilon_2(\omega')}{d\omega'} \ln\frac{1}{|\omega'^2 - \omega^2|}\, d\omega'. \quad (4.14)$$

For the square-root dependence of ϵ_2 we find that $d\epsilon_2/d\omega$ varies as $(\omega-\omega_0)^{-1/2}$ for $\omega>\omega_0$. $\epsilon_1(\omega)$ will therefore also possess a maximum at ω_0 where the weighting function $\ln 1/|\omega'^2-\omega^2|$ for $\omega=\omega_0$ has its maximum. The actual position of the maximum will also be somewhat modified by the frequency dependence of $\omega^{-2}|M_{vc}|^2$ but it will be very close to ω_0 or E_0. Quantitatively at M_0 type critical points the density of states is rather small compared with the densities at higher energies so that k is small compared with n. Consequently both n and R (which in this case is mostly influenced by n) will peak near ω_0. In practice most M_0 edges in low band gap materials appear to be associated with such low densities of states that this maximum in reflectance is not observable. The lead salts PbS, PbSe and PbTe are an exception to this.

No maximum in reflectance is observable for certain other types of absorption threshold. So far we have assumed $|M_{vc}|$ to be constant but different from zero over small intervals corresponding to transitions allowed in the electric dipole approximation. If $M_{vc}=0$ at $\mathbf{k}_0$ the transition is said to be forbidden. One then expands the matrix element near $\mathbf{k}_0$ and takes the first non-zero term which is proportional to $|\mathbf{k}-\mathbf{k}_0|$[53]. Consequently $|M_{vc}|^2$ is proportional to $(E-E_0)$ and $\epsilon_2 \propto (E-E_0)^{3/2}$. For this behaviour the first term in equation (4.14) does not give rise to a peak at E_0, thus the maximum positions of ϵ_1, n and R become rather more unpredictable if in fact a maximum does exist at all. The same is true for indirect transitions which were not included in the treatment in Section 4.1 since only the interaction of electrons with the electromagnetic field was considered. When lattice vibrations are also considered as an additional perturbation, indirect, non-vertical transitions may be allowed in second-order perturbation theory. The change in momentum of the electrons between initial and final states is then furnished by the momentum of the phonon simultaneously absorbed or emitted. The energy dependence of ϵ_2 for allowed transitions is[53]

$$\epsilon_2 \propto (E-E_0)^2 \tag{4.15}$$

and for forbidden transitions

$$\epsilon_2 \propto (E-E_0)^3. \tag{4.16}$$

The absolute values of ϵ_2, however, are some orders of magnitude lower than for direct transitions. Hence indirect transitions can only be observed in absorption at energies where no direct transitions occur and cannot be detected in reflectance.

The behaviour of the optical constants near saddle points is of great importance. As expected from the local character of the dispersion relations, at an M_1-type critical point $\epsilon_1(\omega)$ has a maximum at an energy lower than E_0[54], hence the reflectivity will also peak below E_0. ϵ_2, however, shows a sharp change of slope exactly at E_0. Saddle-point energies should thus be well characterized in the ϵ_2 spectra provided the line shape is caused solely by the interband transition. In practice pure interband line shapes may be obscured by scattering and many body (exciton) effects so that a simple analysis is difficult. On the other hand, such exciton effects may in fact facilitate the location of critical-point energies: the resultant line sharpening combined with the local character of the dispersion relations causes structure in R, ϵ_1 and ϵ_2 to occur very nearly at the same energy. These considerations in most cases enable one to determine critical point energies directly from primary data (such as reflectivity spectra) without introducing errors of more than 0·2 eV. The shift of the critical point energy of M_1 type saddle points characterized by the maximum position of $\epsilon_2\omega^2$, away from the maximum position in R, becomes larger as the measured peak broadens.

4.4. Classification of electronic states

4.4.1. *Spatial symmetry*

So far we have discussed the relation for $\epsilon_2(\omega)$ only qualitatively in terms of the joint density of states and the transition matrix element. For a detailed comparison with the experimental results we require the knowledge of the $E(\mathbf{k})$ dependence of all valence and conduction bands as well as the wave functions associated with the states involved in interband transitions. Both the energy-band structure and the wave functions of the crystal electrons are largely determined by the symmetry of the periodic crystal potential in which the electrons move and thus by the symmetry of the crystal. The classification of electron states in a real crystal must therefore commence with a symmetry analysis of that crystal.

The symmetry of a crystal can be expressed by its space group consisting of all transformations which leave the crystal invariant. These transformations can be translations, rotations, inversions or combinations thereof. The primitive translations form an Abelian subgroup of the space group and their one-dimensional irreducible representations can be characterized by the electron wave vector **k**. The remaining transformations (with the exception of glide reflections and screw displacements) form the point group of the crystal. The Hamiltonian for an electron in the periodic crystal potential is invariant for all transformations of the point group. In this way the symmetry is transferred on to the Hamiltonian so that the irreducible representations of a point group characterize the eigenvalues of the operator and give their degeneracy and the symmetry of the associated wave functions.

The periodic lattice is described by a set of vectors $\mathbf{R}_i$ such that $V(\mathbf{r}+\mathbf{R}_i) = V(\mathbf{r})$ and

$$\mathbf{R}_i = n_{i1}\mathbf{a}_1 + n_{i2}\mathbf{a}_2 + n_{i3}\mathbf{a}_3 \tag{4.17}$$

where the $\mathbf{a}_1$, $\mathbf{a}_2$, $\mathbf{a}_3$ are three independent primitive translation vectors. Now we define a set of vectors $\mathbf{K}_j$:

$$\mathbf{K}_j \cdot \mathbf{R}_i = 2\pi n_{ij}, \tag{4.18}$$

where n_{ij} is an integer. All vectors $\mathbf{K}_j$ then describe the reciprocal lattice with the primitive translation vectors $\mathbf{b}_1$, $\mathbf{b}_2$, $\mathbf{b}_3$

$$\mathbf{K}_j = m_{j1}\mathbf{b}_1 + m_{j2}\mathbf{b}_2 + m_{j3}\mathbf{b}_3. \tag{4.19}$$

The unit cell of the reciprocal lattice is called the Brillouin zone and is found by constructing the planes which bisect the vectors from the origin to all points of the reciprocal lattice. The Brillouin zone is then the figure bounded by those planes which are nearest to the origin. It can easily be shown that the reciprocal lattice is invariant under the same point group as the corresponding real lattice. The groups of a wave vector **k** in the reciprocal lattice consist of all transformations which leave **k** invariant or transform it into an equivalent wave vector $(\mathbf{k}+\mathbf{K}_j)$.

For detailed treatments of the group theoretical formalism the reader is referred to a number of books.[55–8] The elements of a group fall into a number of different classes and one can in addition determine the irreducible representations of the group. The number of irreducible representations is equal to the number of classes. The sets of traces of these representations, written in the form of matrices up to the order 3, form the character table containing all the information relevant to the discussion of the

TABLE 4.2. *Character table of point group* O_h.
Notation according to Bouckaert, Smoluchowski and Wigner (BSW)[59], and Koster *et al.* (K)[60]

BSW		E	$3C_4^2$	$6C_4$	$6C_2$	$8C_3$	J	$3JC_4^2$	$6JC_4$	$6JC_2$	$8JC_3$
	K	E	$3C_2$	$6C_4$	$6C_2^1$	$8C_3$	I	$3\sigma_h$	$6S_4$	$6\sigma_d$	$8S_6$
Γ_1	Γ_1^+	1	1	1	1	1	1	1	1	1	1
Γ_2	Γ_2^+	1	1	−1	−1	1	1	1	−1	−1	1
Γ_{12}	Γ_3^+	2	2	0	0	−1	2	2	0	0	−1
Γ'_{15}	Γ_4^+	3	−1	1	−1	0	3	−1	1	−1	0
Γ'_{25}	Γ_5^+	3	−1	−1	1	0	3	−1	−1	1	0
Γ'_1	Γ_1^-	1	1	1	1	1	−1	−1	−1	−1	−1
Γ'_2	Γ_2^-	1	1	−1	−1	1	−1	−1	1	1	−1
Γ'_{12}	Γ_3^-	2	2	0	0	−1	−2	−2	0	0	1
Γ_{15}	Γ_4^-	3	−1	1	−1	0	−3	1	−1	1	0
Γ_{25}	Γ_5^-	3	−1	−1	1	0	−3	1	1	−1	0

symmetry of the states. The character tables of all thirty-two point groups have been given by Koster *et al.* Table 4.2 shows as an example the character table of the holosymmetric cubic point group. It is the group of the wave vector $\mathbf{k} = 0$ for all cubic materials having inversion symmetry, and thus also for materials with diamond structure. We have ten irreducible representations corresponding to ten possible types of states. The notation in the first line and column is that of Bouckaert, Smoluchowski and Wigner (BSW)[59] and is motivated by the compatibility with states going out from Γ, e.g. (see Fig. 4.3) Γ_{25} with Δ_2 and Δ_5. In the second line and column the notation of Koster *et al.*[60] is given. The characters in the third column, i.e. the traces of the transformation matrices of the identity transformation, give the degeneracies of the states. The set of characters belonging to a given representation gives the symmetry of the associated eigenfunctions. Since the coordinates x, y, z transform according to Γ_{15}

and since *p*-functions are proportional to *x*, *y*, *z*, the state Γ_{15} is of *p*-symmetry. (It should be noted that although this is also the group of the wave vector $\mathbf{k} = 0$ for diamond structures, because of the existence of a second equal atom in the unit cell giving rise to a glide reflection in the space group, the transformation properties are different and consequently the threefold degenerate *p*-state in diamond is Γ'_{25} or Γ_{15}.) Γ_1 and Γ'_2 are of *s*-symmetry and the fivefold degenerate atomic *d*-term splits under the cubic crystal field into the *d*-like states Γ'_{25} and Γ_{12}.

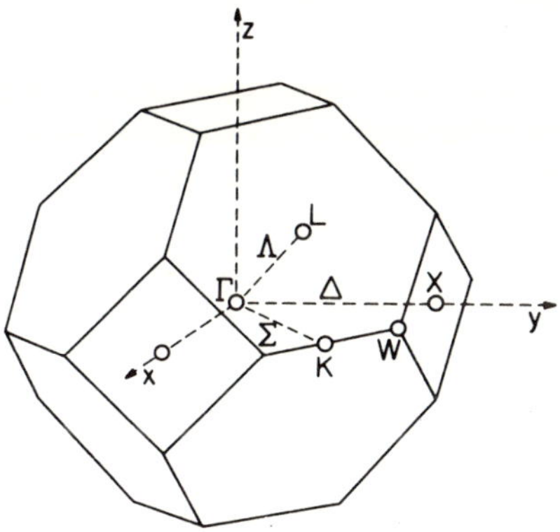

FIG. 4.3. Brillouin zone of the face-centred cubic lattice.

In this way one also classifies the states which can exist at all points where $\mathbf{k} \neq 0$ according to the character table of the groups of the different wave vectors **k**. These groups are in general subgroups of the group of Γ or isomorphic with it. The analysis can, for example, be made for the point *L*, the midpoint of the hexagonal $\langle 111 \rangle$ faces of the Brillouin zone of the face-centred cubic lattice shown in Fig. 4.3 and for a point Λ along the $\langle 111 \rangle$ axis. One finds that there is no state at *L* or Λ of higher degeneracy than 2. The threefold Γ'_{25} state, therefore, splits into different states along the Λ-axis and the compatibility relations tell us which states along the axis can join into the states at high symmetry points. There is an important consequence of the existence of a glide reflection in the space group of the diamond structure connected with the group of the wave vector at *X*. This is the fact that there exist only two-dimensional representations, i.e. we have only twofold degenerate states. Such behaviour is characteristic of space groups containing glide reflections or screw displacements and requires at least two identical atoms per unit cell.

From the symmetry analysis we can obtain a catalogue of states at every point in the Brillouin zone. The final ordering of the levels, however, has to be determined by the choice of the correct crystal potential and a subsequent band structure calculation. From the symmetry we can only determine at which points the condition (4.10) for the occurrence of a critical point is fulfilled, i.e. where all bands have vanishing gradients. For face-centred cubic lattices these are the points Γ, X, L and W.

The zincblende structure is obtained from the diamond structure by replacing the two equal atoms in the diamond unit cell by two dissimilar ones. This means that the inversion symmetry is lost and the appropriate point group is T_d. Its character table is the upper part of Table 4.3 considering only the classes E, $8C_3$, $3C_2$, $6S_4$ and $6\sigma_d$. The threefold degenerate p-state in zincblende is Γ_{15} and equivalent to Γ'_{25} in diamond. Similarly the Γ'_2 diamond state goes into Γ_1. Since there are two different atoms in the unit cell the space group contains no glide reflection. Consequently the "sticking together" of bands at X due to the existence of a glide reflection does not occur and most of the twofold degeneracies at X are lifted.

4.4.2. *Spin–orbit coupling*

In Section 4.4.1 only the orbital part $\phi(r)$ of the electron wave function was considered. If we also want to take spin–orbit coupling, which is essentially a relativistic effect, into account

TABLE 4.3
Character table of double group of T_d.

BSW	K	E	$\bar{E}$	$8C_3$	$8\bar{C}_3$	$3C_2$ $3\bar{C}_2$	$6S_4$	$6\bar{S}_4$	$6\sigma_d$ $6\bar{\sigma}_d$
Γ_1	Γ_1	1	1	1	1	1	1	1	1
Γ_2	Γ_2	1	1	1	1	1	−1	−1	−1
Γ_{12}	Γ_3	2	2	−1	−1	2	0	0	0
Γ_{25}	Γ_4	3	3	0	0	−1	1	1	−1
Γ_{15}	Γ_5	3	3	0	0	−1	−1	−1	1
	Γ_6	2	−2	1	−1	0	$\sqrt{2}$	$-\sqrt{2}$	0
	Γ_7	2	−2	1	−1	0	$-\sqrt{2}$	$\sqrt{2}$	0
	Γ_8	4	−4	−1	1	0	0	0	0

we must include the spin variables by writing $\psi_+ = \phi(\mathbf{r})u_+$ and $\psi_- = \phi(\mathbf{r})u_-$. The new Hamiltonian containing the spin–orbit term is only invariant under the transformations applied to both the orbital and the spin variables simultaneously. The number of operations of a symmetry group is doubled since the spin functions u_+ and u_- change sign under a rotation by 360°, denoted $\bar{E}$, corresponding to a transformation of the spin variables. The number of classes is not necessarily twice the number of classes of the single group. It is found that new irreducible representations are needed to describe the states of the double group. The character table of the double group belonging to T_d is shown in Table 4.3 as an example. We give both the BSW notation[(59)] and the notation used by Koster *et al.*[(60)] Now Γ_6 is the representation formed by the spin functions and the states of the single group go into those of the double group by multiplication with Γ_6. The new states can easily be found by character multiplication and it is found in this case that $\Gamma_1 \rightarrow \Gamma_6$, $\Gamma_2 \rightarrow \Gamma_7$ and $\Gamma_{12}(\Gamma_3) \rightarrow \Gamma_8$. The states $\Gamma_{25}(\Gamma_4)$ and $\Gamma_{15}(\Gamma_5)$ become sixfold degenerate by the multiplication and since there is no state of such degeneracy in the double group they will split. $\Gamma_{15}(\Gamma_5)$ goes into $\Gamma_7 + \Gamma_8$ and $\Gamma_{25}(\Gamma_4) \rightarrow \Gamma_6 + \Gamma_8$. This means that the Γ_{15} state, threefold degenerate in the orbital group and of *p*-symmetry in zincblende crystals, splits into a fourfold state Γ_8 [often called $\Gamma_{15}(\frac{3}{2})$] and a twofold state Γ_7 [or $\Gamma_{15}(\frac{1}{2})$]. The same splitting occurs for the Γ'_{25} state in diamond but we can picture the split bands as of degeneracy 2 and 1 as a consequence of the space and time degeneracy taking place in crystals with inversion symmetry. For zincblende crystals there is also an energy term linear in $|\mathbf{k}|$ so that the band extrema occur off $\mathbf{k} = 0$ but these are very small effects. The vertical splitting, however, is expected to be of the same order of magnitude as in the free atoms and spin-orbit splittings of *p*-states indeed play a very important role in the interpretation of the fundamental optical spectra.

4.5. Methods of band structure calculation

4.5.1. *General considerations*

The joint density of states of valence and conduction bands which is the most important factor determining ϵ_2 in equation (4.7) can only be derived from a detailed knowledge of the form of the energy bands $E_v(\mathbf{k})$ and $E_c(\mathbf{k})$ in all symmetry directions of **k**-space. The symmetry considerations give much qualitative information and can simplify band structure calculations considerably. The quantitative dependence of $E(\mathbf{k})$, however, can only be found from a solution of the Schroedinger equation in one way or another. There exist a great number of approximate methods[61] and we will describe some of them very briefly. Common to all is the use of the Born–Oppenheimer approximation, i.e. the motion of the nuclei is separated from that of the electrons, and the electron states are defined under the assumption that the nuclei are fixed in their actual positions. Furthermore, all methods assume the one-electron approximation, i.e. one calculates the wave function of one electron in the field formed by the lattice potential and by all other electrons. The lattice potential is constructed as a sum of atomic potentials at all lattice points. The variety of methods differs in the choice of the potential, the superposition of the local potentials, and the technique used to solve the Schroedinger equation. The two main groups are those within the tight binding approximation and those based on the free electron approach. The former ones are well suited if the overlap between the wave functions associated with the different lattice sites is small, e.g. for the calculation of the valence bands of the alkali halides. The free electron approximation is the other extreme and is, e.g., suitable for the alkali metals.

Finally, the third feature common to all methods is that one expands the unknown wave functions in a set of known wave functions, e.g. into plane waves of the form

$$\psi_{\mathbf{K}} = \sum_j a_{\mathbf{K}_j} e^{j(\mathbf{k}+\mathbf{K}_j)\cdot\mathbf{r}} . \tag{4.20}$$

It is advantageous to use a set of symmetrized linear combinations of plane waves which transform according to the irreducible representations of the group of wave vector **k**. In this way one works with wave functions which have the required symmetry built in.

4.5.2. *Free electron model*

In this approximation, also called the empty-lattice method, the periodic potential is regarded as arbitrarily weak so that it can be neglected in the Hamiltonian. By using symmetrized linear combinations of plane waves in the expansion one incorporates the crystal symmetry. The energy is then simply

$$E = \frac{\hbar^2}{2m} |\mathbf{k} + \mathbf{K}_j|^2 \tag{4.21}$$

and since it depends only on **k** and the reciprocal lattice vectors, the $E(\mathbf{k})$ dependence is the same for all crystals of a certain crystal class. Figure 4.4 shows as an example the free electron bands along the $\langle 111 \rangle$ axis for a face-centred cubic lattice. Equation (4.21) gives a parabola at each point of the reciprocal

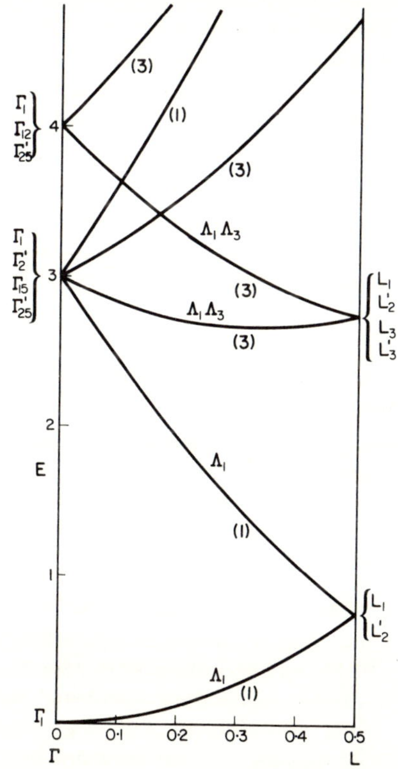

FIG. 4.4. Energy bands in the free electron limit as a function of wave vector for points along the $\langle 111 \rangle$ axis of the f.c.c. lattice Brillouin zone (Jones[55]). Energies are given in units of $2\hbar^2\pi^2/ma^2$.

lattice and the parts of the parabolas lying within the Brillouin zone then form the $E(\mathbf{k})$ structure shown in Fig. 4.4. In order to determine the position of the Fermi surface or of a possible band gap one fills up the energy bands with the available electrons. Because of the high degeneracy of states in this model, e.g. eightfold at $E_\Gamma = 3$ in Fig. 4.4, it is advisable to go one step further to the nearly free electron approximation in which the periodic potential $V(\mathbf{r})$ is introduced as a small perturbation. $V(\mathbf{r})$ can be expanded in a Fourier series

$$V(\mathbf{r}) = \sum_j V_{\mathbf{K}_j} e^{i\mathbf{K}_j \cdot \mathbf{r}}$$

since plane waves of the type $e^{i\mathbf{K}_j \cdot \mathbf{r}}$ have the full translational periodicity of the potential. Under this perturbation many of the degeneracies of the free electron model are lifted and the splittings can be calculated by solving a secular determinant. If E is twofold degenerate because of, e.g., $|\mathbf{k}+\mathbf{K}_0|^2 = |\mathbf{k}+\mathbf{K}_1|^2$, the energy values split by the amount $2|V_{\mathbf{K}_1-\mathbf{K}_0}|$. If the Fourier coefficient $V_{\mathbf{K}_1-\mathbf{K}_0}$ vanishes the degeneracy is not lifted in this approximation. This method has been used extensively in constructing the Fermi surface of metals(62).

4.5.3. *OPW and pseudopotential method*

Since one is interested particularly in the states of the valence electrons, i.e. the states of the highest valence and the lowest conduction bands, one can expand into plane wave functions which are orthogonal to the wave functions of the core electrons. These functions are therefore called orthogonalized plane waves(63) and have the form

$$\psi_{\mathbf{K},\mathbf{k}} = e^{i(\mathbf{k}+\mathbf{K}) \cdot \mathbf{r}} - \sum_i b_{\mathbf{k},\mathbf{K}_i} \phi_{\mathbf{K}_i}(\mathbf{r})$$

where $\phi_{\mathbf{K}_i}$ are the core wave functions. The $\phi_{\mathbf{K}_i}$ are in practice approximated by Bloch sums over atom-like functions obtained from a tight-binding calculation. The $\psi_{\mathbf{K},\mathbf{k}}$ then behave like a plane wave at large distances from an atom but have the rapidly varying character of an atomic wave function nearby. With an expansion of the unknown wave functions into the $\psi_{\mathbf{K},\mathbf{k}}$'s one obtains the lowest valence band state as the lowest solution of the secular determinant. For good convergence properties of the eigenvalues it is still necessary to take a large number of plane waves for the expansion.[64]

In the pseudopotential method[65] a repulsive potential V_R is added to the Hamiltonian

$$H_0 = \frac{\mathbf{p}^2}{2m} + V_c(\mathbf{r})$$

so that the lowest eigenvalue of $(H_0 + V_R)$ is also the lowest valence band state. The repulsive potential is chosen in such a way that the rapidly varying character of the wave function near the nuclei is largely compensated. One therefore requires very few Fourier coefficients of the potential in the expansion and hence the calculation is much simplified. Using this procedure it becomes possible to calculate $E(\mathbf{k})$ at many points of the Brillouin zone (e.g. in Ge[66] with three remaining Fourier coefficients used as adjustable parameters) so that a dense mapping of the joint density of states can be made.

4.6. Selection rules

The interband contribution to ϵ_2 is determined by the values of both J_{vc}, the joint density of states, and the matrix element M_{vc} defined by equation (4.5). As we pointed out in Section 4.2 M_{vc} is a slowly varying function of $\mathbf{k}$. With the selection rule for direct transitions $\mathbf{k}' = \mathbf{k} + \boldsymbol{\kappa}$ we can write

$$M_{vc} = \int_V \mathbf{dr}\, \psi_c^*(\mathbf{k},\mathbf{r})\, \mathbf{e} \cdot \boldsymbol{\nabla}\, \psi_v(\mathbf{k},\mathbf{r}), \tag{4.22}$$

where the ψ_c and ψ_v are Bloch wave functions. According to Seitz[67] this is equivalent to

$$M_{vc} = \left[\int_V \mathbf{dr}\, \psi_c^*\, \mathbf{e} \cdot \mathbf{r} \psi_v\right]\left(-\frac{m}{\hbar^2}\right)(E_c - E_v),$$

where $\mathbf{e} \cdot \mathbf{r}$ is the coordinate in the direction of the electric vector of the light. For light polarized in the x, y or z direction $\mathbf{e} \cdot \mathbf{r}$ is proportional to x, y or z, respectively. In most cases the wave functions are not explicitly known. From the character tables, however, we know the symmetry properties of the wave functions. This enables us to calculate the selection rules, i.e. to determine whether M_{vc} has zero or finite values. From the orthogonality of the functions belonging to the different rows of an irreducible representation or to different irreducible representations, one obtains[49] the following rule governing the transition from a state i to a state j

$$C_i^j = \frac{1}{h} \sum_R \chi_j(R) \chi_{x,y,z}(R) \chi_i(R) \begin{cases} = 0 \text{ forbidden} \\ \neq 0 \text{ allowed} \end{cases} \quad (4.23)$$

where $\chi_j(R)$, $\chi_i(R)$ are the characters of the jth or ith representation, respectively, of the transformation R. $\chi_{x,y,z}(R)$ are the characters of the irreducible representation according to which the coordinates x, y, z transform. h is the number of all transformations R in the group. By summing the products of the χ's one can thus determine if a certain transition is either allowed or forbidden by symmetry considerations. In crystals of cubic symmetry x, y, z transform according to the same irreducible representation; the matrix element is thus independent of the direction of polarization. By inspection of the character table of the point group 0_h (Table 4.2) where the coordinates belong to Γ_{15} we find, for example, that electric dipole transitions from Γ'_{25} to Γ'_2, Γ'_{12}, Γ_{15} and Γ_{25} states are allowed and transitions from Γ'_{25} to all other states are forbidden.

In crystals of non-cubic symmetry x, y and z will transform according to different irreducible representations. The selection rules will thus depend on the direction of polarization of the light. As an example we will discuss the selection rules for the point group D_{3d} of rhombohedral symmetry in Chapter 6.

CHAPTER 5

FUNDAMENTAL OPTICAL SPECTRA OF CUBIC SEMICONDUCTORS AND INSULATORS

5.1. General

In this chapter we shall be concerned with the interpretation of the fundamental spectra of cubic semiconducting and insulating materials. Most of the spectra that we analyse are the results of near-normal incidence reflectivity measurements using essentially unpolarized radiation. As we pointed out in Chapter 2 a number of methods exist for determining the optical constants (measurement at oblique incidence, etc.). Experimentally it has been found that normal incidence reflectivity represents the most satisfactory and convenient method when a wide energy range is considered. The correspondence between the optical constants for semiconductors from normal incidence reflectivity, and an ellipsometric method, has indeed been demonstrated by Marple and Ehrenreich for Ge and CdTe.[(68)] Although in principle one obtains the same information from either the fundamental absorption spectrum or the reflectivity spectrum, the difficulty in preparing thin specimens for measurement of absorption coefficients of up to around $10^6\ cm^{-1}$, which have the properties of the bulk material, has caused a marked sparsity of absorption measurements in this field. This applies particularly to the diamond and zincblende materials. This is not the case for more ionic cubic materials such as the alkali halides, where it has been recognized for many years that reliable thin films can be prepared by very simple techniques.

By fundamental region, we mean the energy range from the lowest absorption edge up to an energy where no more structure is observable in the optical constants of the material. For almost all the materials we consider (with, perhaps, the exception of some alkali halides), little structure is observable beyond about

20 eV and hence rather complete data on the optical constants is available using straightforward spectroscopic techniques.

The discussion of cubic materials given here centres mainly around the diamond and zincblende semiconductors, and materials having the rocksalt structure (e.g. PbS, PbSe, PbTe and the alkali halides). We also deal briefly with closely related materials such as the group V semimetals whose band structure can be regarded as a perturbation of the cubic case. The band structure of the wurtzite compounds is dealt with separately in Chapter 6.

5.2. Direct interband transitions and van Hove singularities

We have seen in Chapter 2 that the knowledge of either the absorption coefficient or the reflectivity at normal incidence over the fundamental region, together with one of the Kramers–Kronig relations and a reasonable high-energy extrapolation, enables us to determine the remaining optical constants and the real and imaginary parts of the dielectric constants ϵ_1 and ϵ_2. If we can assume that the matrix element for the interband transitions we are considering remains approximately constant, then we can also determine the shape of the density of states function, as this should be proportional to $\epsilon_2 E^2$, where E is the photon energy. Experimentally it is found that coarse structure (e.g. peaks) in the reflectivity spectrum manifests itself as peaks in the absorption coefficient, ϵ_2 and $\epsilon_2 E^2$. The integral transform of the Kramers–Kronig relation may, however, shift structures in ϵ_2 and $\epsilon_2 E^2$ by small amounts. Thus although it may be sufficient to use energies obtained directly from reflectivity data for systematic studies on a series of materials, comparison with theory should be made with energies obtained from ϵ_2 curves. This is particularly important when studying the line shape of a transition. As was demonstrated in Chapter 4, the line shape in principle allows the nature of the transition to be determined. The line shape of a transition observed in the reflectivity or absorption spectrum may not be reproduced accurately in the ϵ_2 curve; there will, however, in general be sufficient identity between the two curves to use the direct experimental data to recognize the presence of any critical points. The behaviour of ϵ_2 around critical points shown in Fig. 4.1 leads to another important reiteration: a single critical point or singularity in the density of states does not in itself produce a peak in ϵ_2 but only an edge. To obtain a peak due solely to interband transition, there must be an accidental degeneracy of M_1 and M_2 singularities. It is such a degeneracy that gives rise to the main reflectivity peak at 4·5 eV in germanium and silicon.

The analysis of the fundamental optical constants of semiconductors in terms of van Hove singularities has been one of the main factors leading to a comparatively recent deeper understanding of the band structure of these materials.[69] Unfortunately, in the diamond and zincblende semiconductors clear examples of singularities which can be fitted directly to the analytic behaviour shown in Fig. 4.1 are rare. As we see in the following sections, such examples that do exist are best revealed by low-temperature measurements, and are obscured at room temperature. The fact that in all the common semiconductors and insulators the fundamental optical constants are dominated by peaks and not by edges implies that additional mechanisms are involved. It is now realized that these peaks are mostly caused by critical point transitions accompanied by strong electron-hole interaction (see Chapter 8). Again, low-temperature measurements are useful in identifying these exciton or exciton-like effects, as the line narrowing which occurs when the temperature is reduced is considerably greater than one would predict from interband effects alone.

Historically for diamond and zincblende semiconductors, the identification of the spin-orbit splitting of the top valence band provided the first clue to the evaluation of the deep-lying band structure.[70] In germanium, for example, the splitting at the zone centre can be found from infrared intraband absorption measurements, and one can predict the value of two-thirds of this in the $\langle 111 \rangle$ direction that is observed in the reflection spectrum. A search for the presence of spin–orbit effects, particularly in heavy materials where one expects substantial splittings, still represents a good approach to the evaluation of the band structure of a new material.

5.3. Group IV semiconductors

We deal mainly with germanium and silicon, and only briefly with the other group IV covalent materials diamond and grey tin. The interband reflectivity spectra of germanium and silicon were first investigated by Phillip and Taft[11,71] who showed the presence of two large reflectivity maxima, the main one located at 4·5 eV. Figure 5.1 shows the room-temperature reflectivity spectra of germanium and silicon. The splitting of the low-energy peak in Ge was first observed by Tauc and Antončik[70] and assigned to valence band spin–orbit splitting probably at the L point of the Brillouin zone. In fact the similarity between the low-energy peaks of germanium and silicon is misleading; in Ge it is now clear that the E_1 peak is caused by M_1 saddle-point transitions in the $\langle 111 \rangle$ direction. In Si, however, two pieces of experimental evidence point to a different assignment: Tauc and Abraham[72] measured the optical properties of Ge – Si alloys and clearly showed a break at approximately 75 atomic per cent Si for the E_1 peak, which indicated a different origin in the two elements. Secondly,

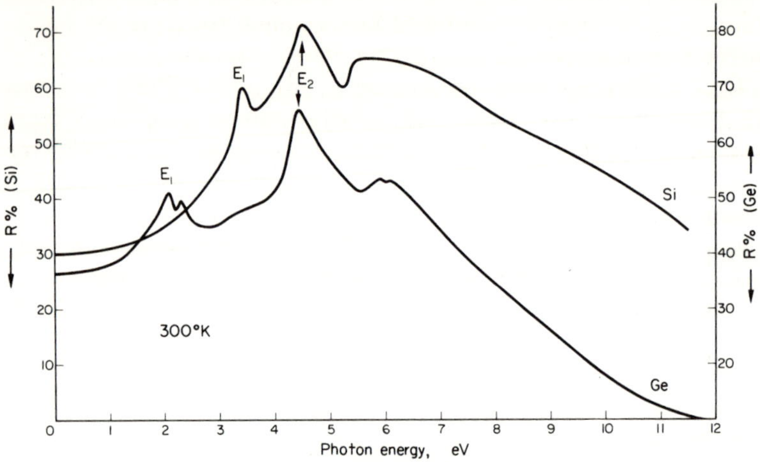

FIG. 5.1. Room temperature reflectivity of silicon and germanium.

it is found that the two peaks have quite different temperature coefficients: about -4×10^{-4} eV/°C for the 2·1–2·3 eV doublet in Ge, and -2×10^{-4} eV/°C for the 3·4 eV E_1 single peak in Si. Historically, the large 4·5 eV peaks were assigned to transitions near the $\langle 100 \rangle$ zone boundary (X-point) and experimental confirmation of this came from low-temperature reflectivity data on zincblende semiconductors (see following section) where the expected splitting of the X conduction band caused by the removal of inversion symmetry was observed. The first major theoretical support for the wealth of experimental data which was accumulating on the diamond (and zincblende) materials came from the elegant pseudopotential energy band calculations performed first on Ge and later on Si by Brust, Bassani and Phillips.[66,73,74] It was found that just three parameters gave an excellent fit to all the important valence and conduction band levels contributing to the optical spectra. Fig. 5.2 shows the comparison of ϵ_2 derived from the pseudopotential calculation and from reflectivity data for Ge. The results showed that the interpretation of the reflectivity data was essentially correct but incomplete. The important result emerged that additional van Hove singularities were required in order to explain the line shape of the major peaks and edges. In germanium, the 2·1, 2·3 eV E_1 doublet was found to result from M_1 edges due to Λ_3–Λ_1 transitions along the $\langle 111 \rangle$ direction. The L_3'–L_1 (M_0) thresholds to which the doublet had previously been assigned were predicted to occur at 1·9 and 2·1 eV and to be much weaker. These weak transitions are difficult to detect in Ge[75], but, as we shall see later, have been clearly observed in some zincblende materials. The 4·5 eV E_2 peak arises from the accidental de-

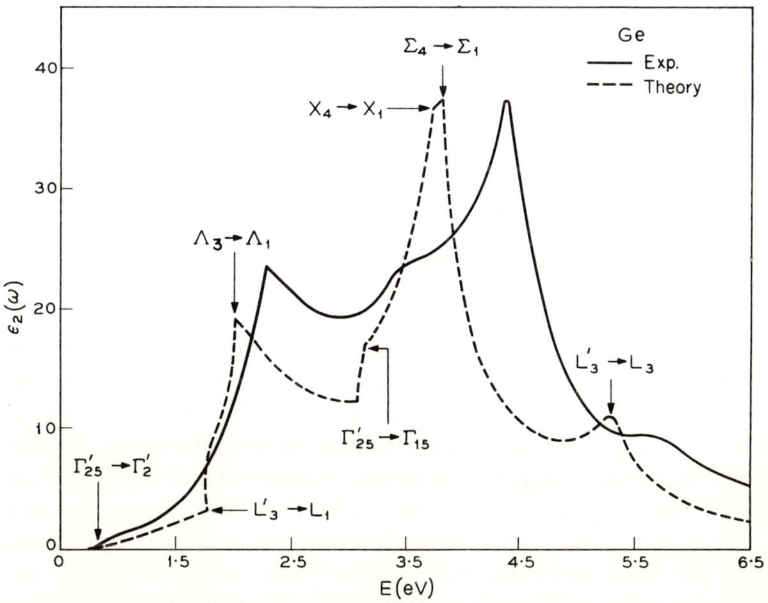

FIG. 5.2. Imaginary part of the dielectric constant $\epsilon_2(\omega)$ for Ge. The solid curve is taken from reflectance data and the dashed curve from the pseudopotential calculations of Brust, Bassani and Phillips.[66]

generacy of an M_1 edge due to X_4–X_1 transitions and an M_2 edge due to Σ_4–Σ_1 transitions. As we shall see in Chapter 8 there is evidence that both the E_1 and E_2 maxima exhibit exciton-like effects which cause line narrowing. The calculation predicts that E_2 in silicon has the same origin as in germanium. The 3·4 eV E_1 peak was assigned to Γ'_{25}–Γ_{15} transitions which complimented the Ge–Si alloy measurements and the temperature coefficient data. Recent work has indicated, however, that it is by no means certain that this interpretation is correct. There is now evidence for contributions to the 3·4 eV peak from $\langle 100 \rangle$ saddle-point transitions near Γ'_{25}–Γ_{15}, from L'_3–L_1 transitions, and also from Γ'_{25}–Γ'_2 transitions. Electroreflectance data (see Section 9.2) clearly indicate a rather complicated degeneracy around 3·4 eV (the Brust *et al.* calculation also indicated accidental degeneracy between Γ'_{25}–Γ_{15} and L'_3–L_1) but a definite assignment for the dominant transition cannot yet be made. It may well be that in addition to well-defined critical points, extended regions of the Brillouin zone contribute to the E_1 peak. Evidence which supports this viewpoint comes from a recent band structure calculation on Si by Kane[76] using a Heine–Abarenkov model potential. In this calculation, the Brillouin zone is divided into a cubic mesh of boxes with side (0·25) $2\pi/a$, and the contribution to ϵ_2 from each box determined. It is seen that near 3·4 eV, ϵ_2 has sizeable contributions from the centre box (Γ) and from boxes near Γ in all major symmetry directions. The major contribution arises from states in the $\langle 100 \rangle$ direction. The 4·5 and 5·5 eV regions are seen also from this model to possess a compound nature. On the basis of these results, it thus seems well justified to talk of extended critical regions instead of isolated critical points.

Another detailed study of the energy bands of silicon has recently been made by Herman *et al.*[77] This approach starts with a non-relativistic, self-consistent band calculation. The results of this first step are subsequently adjusted to reliably established features of the bands by including a small empirical correction. The final results indicate—in agreement with Kane—that the 4·5 eV peak is only to a small extent caused by X_4–X_1 transitions and that the Γ'_{25}–Γ_{15} transitions (also in Ge and α-Sn) are lower than previous estimates by at least 0·5 eV, namely at 2·8 eV in Si, 2·7 eV in Ge and 2·2 eV in α-Sn. Until now there is no experimental evidence concerning these latter results.

Above the E_2 peak, the reflectivity spectra reveal peaks at 5·5 eV in Si and at 5·9 and 6·1 eV in Ge which are attributed to L_3'–L_3 transitions. These are also revealed by the Brust *et al.*[(66)] calculation, although the predicted magnitudes of the transitions are considerably greater than those observed experimentally. The assignment seems definite however, as the characteristic 0·2 eV valence band spin orbit splitting is clearly seen in the curve for germanium.

The collected values for the various interband transitions, both experimental and calculated, are given in Table 5.1; this table also includes data for the two other group IV materials diamond and grey tin. Diamond and α-tin are less completely known. Reflectivity data obtained for diamond[(78)] and also for α-Sn[(79)] indicate that the overall band structures show similarities to those of germanium or silicon although diamond has a fundamental band gap of 5·5 eV while α-Sn is a semimetal with zero band gap[(80)]. In particular α-Sn, by virtue of its high atomic number, exhibits a large spin orbit splitting (0·47 eV) which is clearly seen in the E_1 reflectivity peak. Band structure calculations exist for both diamond[(81)] and α-Sn[(82)]; they are not based on the pseudopotential approach, but have been made from first principles. Both calculations have large discrepancies when compared with the available experimental data and indicate the extreme difficulties which are encountered in a basic calculation with no adjustable parameters. Recent pseudopotential calculations on diamond[(83)] will be discussed in Section 8.2.

One further approach to band-structure determination should be mentioned here. Cardona and Pollak[(84)] have used the $\mathbf{k} \cdot \mathbf{p}$ method in combination with experimental data from cyclotron resonance and optical studies to determine the energy bands of Ge and Si throughout the entire Brillouin zone. Excellent agreement is obtained both with the pseudopotential results and with experimental energy gaps not implicitly fitted by the calculation. This general method includes spin–orbit interactions and is valuable for materials with large relativistic effects as these corrections are incorporated into the energy gaps and band parameters at $\mathbf{k} = 0$, which are readily determined by fitting experimental data. Cardona and Pollak's results for Ge and Si are shown in Fig. 5.3.

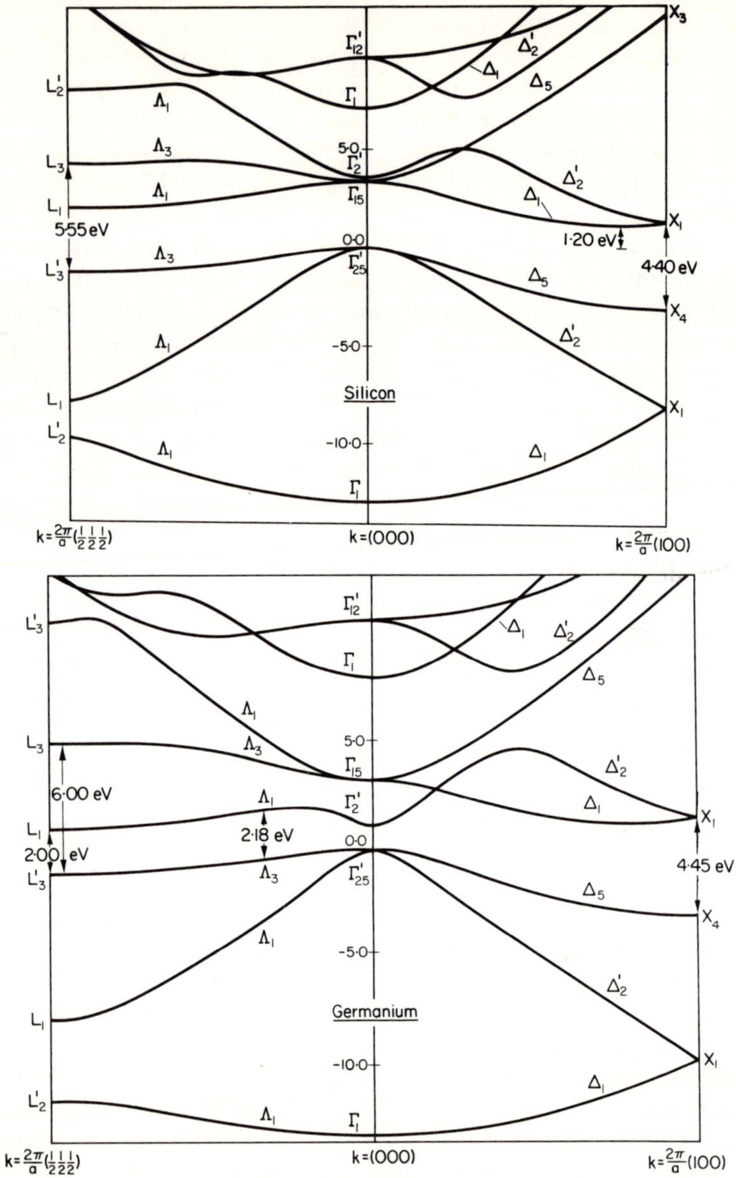

FIG. 5.3. Band structure of Si and Ge in the ⟨111⟩ and ⟨100⟩ directions obtained using the **k** · **p** method (Cardona and Pollak[84]).

TABLE 5.1.

Energy gaps of group IV materials

All gaps are expressed in electronvolts and are measured at room temperature unless otherwise stated.

		Γ'_{25}–Γ'_2	L'_3–L_1	Λ_3–Λ_1	Γ'_{25}–Γ_{15}	X_4–X_1 Σ_4–Σ_1	L'_3–L_3	Optical gap (indirect)
C	Theory		13·6		7·4	13·7		$\Gamma'_{25}-\Delta_1 = 6{\cdot}7$
	Expt.		16·3		7·2	12·6		5·5
Si	Theory		3·15		2·8 or 3·5	4·0 4·4	5·2	
	Expt.	E_1 maximum at 3·4 eV (See text)				4·5	5·5	$\Gamma'_{25}-X_1 = 1{\cdot}1$
Ge	Theory	0·8	1·8	2·0	2·7 or 3·6	3·6 3·8	5·4	
	Expt.	0·8 1·09	1·74 1·94	2·10 2·29	3·2 3·4	4·45	5·9 6·1	$\Gamma'_{25}-L_1 = 0{\cdot}65$
α-Sn	Theory	−0·16	1·4		2·2 or 3·0	3·4	4·4	
	Expt.†			1·28 1·75	2·8	3·5	4·4	

†Reflectance measurement at 200°K.

TABLE 5.2

Abbreviated periodic table

	I		*II*		*III*		*IV*		*V*		*VI*		*VII*		*VIII*		
	A	B	A	B	A	B	A	B	A	B	A	B	A	B			
Outer electrons	1*s*	1*s*	2*s*	2*s*	2*s*,1*p*	1*d*,2*s*	2*s*,2*p*	2*d*,2*s*	2*s*,3*p*	3*d*,2*s*	2*s*,4*p*	4*d*,2*s*	2*s*,5*p*	5*d*,2*s*			
1	H																
2	Li		Be		B		C		N		O		F				
3	Na		Mg		Al		Si		P		S		Cl				
4	K	Cu	Ca	Zn	Ga	Sc	Ge	Ti	As	V	Se	Cr†	Br	Mn	Fe	Co	Ni
5	Rb	Ag	Sr	Cd	In	Y	Sn	Zr	Sb	Nb‡	Te	Mo†	I	Tc§	Ru	Rh	Pd
6	Cs	Au	Ba	Hg	Tl	La*	Pb	Hf	Bi	Ta	Po	W	At	Re	Os	Ir	Pt

*Rare earth series. †5*d*,1*s*. ‡4*d*,1*s*. §6*d*,1*s*.

We have so far considered only the band structure of the top valence band and the conduction bands. For the diamond materials transitions involving these bands give rise to all the structure in the optical constants at readily accessible photon energies. At higher energies one expects structure in the optical reflectivity due to excitation from atomic *d*-levels into the conduction band. Diamond and Si possess no *d*-bands, but the expected weak structure is seen at about 30 eV in Ge[85]. No experimental data are available for α-Sn. As we see in the following section, in the more ionic zincblende semiconductors, the cation *d*-band excitations move nearer the top of the valence band, and eventually overlap the fundamental interband spectrum.

5.4. Zincblende semiconductors

The diamond crystal structure consists of two interpenetrating face-centred cubic lattices intersecting at one-quarter the distance along a body diagonal; there are two atoms in the unit cell. In the zincblende structure the two interpenetrating lattices are of dissimilar atoms from the IIIrd and Vth groups, IInd and VIth groups, or Ist and VIIth groups of the periodic table. The two crystal structures are equivalent except that the inversion symmetry of diamond is absent in the zincblende materials; one expects marked similarities in the band structure as the total electronic configuration is the same for all the materials in both structures, and only the relative amount of charge on each atom differs. Some zincblende materials (e.g. CdS) also crystallize in the optically anisotropic but closely allied wurtzite structure which consists of two interpenetrating close-packed hexagonal lattices; again similarities in the band structure are expected and observed (see Chapter 6). Table 5.2 shows an abbreviated periodic table of the elements in which all the zincblende and other common semiconducting materials are shown. It will be useful to refer to this table in the sections which follow, particularly when we deal with the systematic variation of the band structure for sequences of materials. The similarity of the fundamental band structure for zincblende semiconductors was clearly shown by Tauc and Abraham[86] who measured the reflectivity between the band edge and about 5 eV of several III V compounds. In these spectra, the E_1 and E_2 maxima seen in Ge are evident and the splitting of the E_1 peak due to spin–orbit interaction is also present.

The assignment of the E_2 reflectivity peak to transitions near the X point of the Brillouin zone was, as we have already seen, confirmed theoretically by Brust *et al.* for Ge. At the same time, experimental support for this assignment came from low-temperature (77°K) reflectivity measurements on GaAs, GaSb, InAs and InSb.[87] The removal of inversion symmetry on going from the group IV to the group III V semiconductors causes the removal of the double degeneracy of the X_1 and X_4 energy bands and while the spin–orbit splitting of the top valence band is expected to be small, a splitting of a few tenths of an electron volt is expected for the lowest X_1 conduction band.[88] This splitting was observed in ref. 87 for the four zincblende materials but not for Ge and Si. The low-temperature measurements also revealed additional weak structure in the reflectivity between the E_1 and E_2 maxima which could be attributed to $\langle 000 \rangle$ transitions from the top of the valence band into a higher conduction band ($\Gamma'_{25}-\Gamma_{15}$).† For GaAs and InSb a doublet structure was observed with the expected spin–orbit splitting of the top valence band at the centre of the zone.

The II VI materials ZnTe, CdTe and HgTe provide clear confirmation[89] of Brust's pseudopotential calculation which assigns the E_1 doublets to Λ_3–Λ_1 transitions. The reflectivity spectrum of ZnTe at room temperature and 77°K is shown in Fig. 5.4; these results show the emergence at low temperatures of the weak M_0 thresholds (just below the main E_1 doublet components) which correspond to the spin orbit split L point transitions. It is seen from this that there is a clear identity between the band structure of the diamond and zincblende structure semiconductors, and in fact the optical measurements on the compounds provide the strongest support for predictions made in the elemental semiconductors. The majority of zincblende materials follow the behaviour of Ge in which the lowest two interband thresholds are associated respectively with Γ'_{25}–Γ'_2 and L'_3–L_1 transitions. GaP exhibits behaviour that parallels Si rather than Ge: as in Si, the indirect optical gap (2·2 eV) corresponds to Γ'_{25}–X_1 transitions. Also the lowest

†In this section we use primarily the diamond notation to illustrate the close similarity between the group IV and the zincblende materials. The correct zincblende notation is shown in Fig. 5.6. Historically, the diamond structure notation has frequently been used to describe the energy states in zincblende.

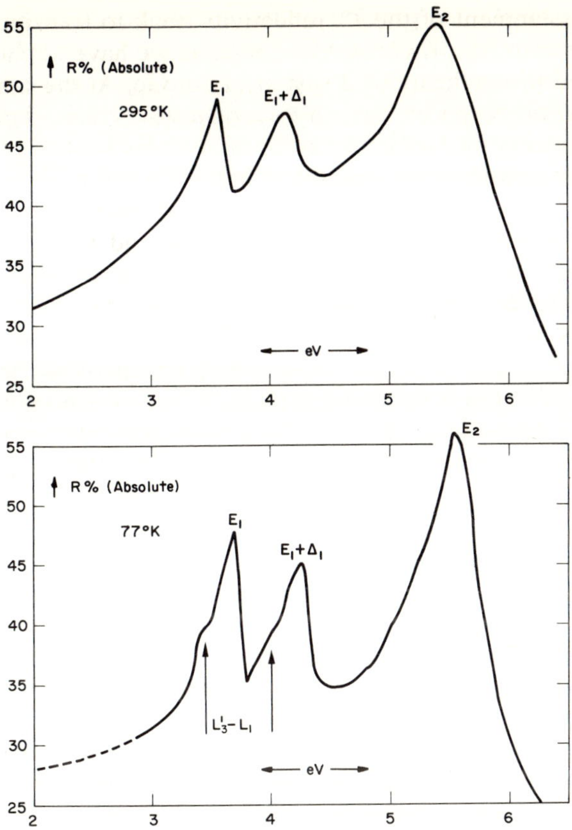

FIG. 5.4. Reflectivity of cleaved ZnTe at room temperature and 77°K (Cardona and Greenaway[89]).

direct threshold Γ'_{25}–Γ'_2 at 2·8 eV is nearer to the value in Si than in Ge. The first strong peak in the reflectivity of GaP lies at 3·7 eV; this peak also exhibits much lower temperature and pressure coefficients[90,91] than those found for Λ_3–Λ_1 transitions in other zincblende materials. Historically, this also led to an assignment to Γ'_{25}–Γ_{15}. Recent electroreflectance data by Thompson *et al.*,[92] however, clearly show that the Γ'_{25}–Γ_{15} transition in GaP lies near 4·8 eV, thus a definite assignment for the dominant transitions causing the 3·7 eV peak cannot be made at present. Since GaP is isoelectronic with Si, it is quite likely that structure in this energy range is again of composite origin.

As we mentioned previously, the energy bands at the X point which are doubly degenerate for the diamond lattice split in zincblende. The small expected splitting of the top valence band (X_5) has been observed by very low-temperature (12°K) reflectivity measurements on HgTe and HgSe.[93] The HgTe curve is shown in Fig.5.5; a number of general assignments can be made for the high-energy structure. Around 8 eV a doublet is seen (with the characteristic $\langle 111 \rangle$ spin–orbit splitting) which is assigned to L_3'–L_3. This structure appears above the E_2 maximum in most of the zincblende materials, as well as in Si and Ge. We note that the X_1–X_3 splitting in HgTe is 0·45 eV whereas the value of this splitting in HgSe is 0·75 eV (see following section). According to Phillips[94] the reflectivity peak at 9·7 eV in HgTe is caused by L_3'–L_2' transitions, and not by atomic d-band excitation as suggested previously;[89] this is supported by the fact that the

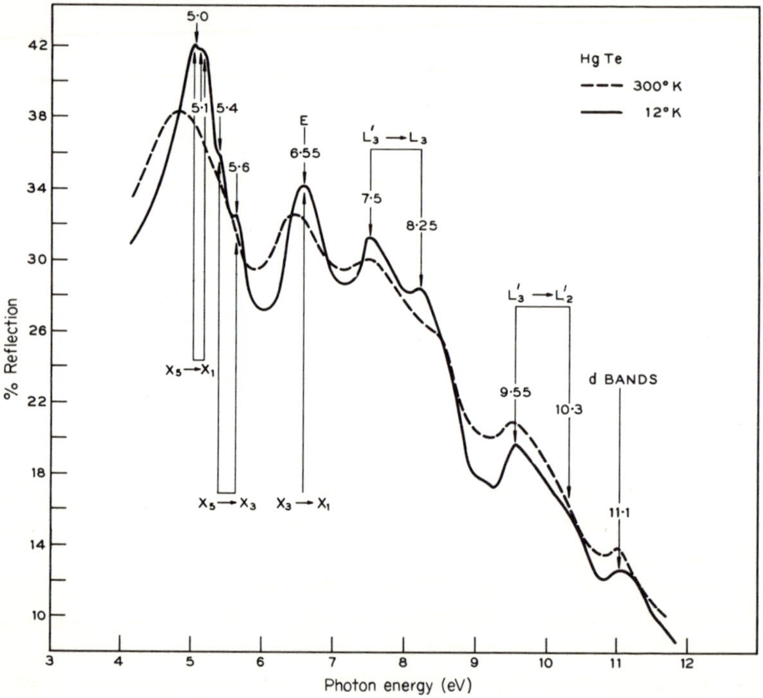

FIG. 5.5. Reflectivity of HgTe at 300°K and 77°K (Scouler and Wright[93]).

TABLE 5.3

Energies (in electron volts) of the direct interband transitions in zincblende semiconductors

The values shown are taken from reflectance or transmission data at room temperature unless otherwise stated (R or T), from structure in the imaginary part of the dielectric constant computed using reflectance data (ϵ_2), and from the pseudopotential calculations of Cohen and Bergstresser[96] (CB). The CB results do not include spin–orbit splitting and refer to $T = 0°$K. It should also be noted that the CB calculations give no information about the Λ_3–Λ_1 critical point transitions which are expected to lie at slightly higher energies than the L_3–L_1 transitions. The experimental data in this table were obtained from the following sources: (a) refs, 12, 92, 145 and Ehrenreich, H., Philipp, H. R. and Phillips, J. C. *Phys. Rev. Letters* **59**, 59–61 (1962) (referred to as EPP); (b) ref. 69; (c) refs. 87, 101, and EPP; (d) ref. 101; (e) refs. 87, 170, and EPP; (f) ref. 89; (g) ref. 101; (h) ref. 145, and Fischer, T. E. *Phys. Rev.* **139A**, 1228–33 (1965); (i) ref. 120; (j) Aven, M., Marple, D. T. F. and Segall, B., *J. Appl. Phys.* **32**, 2261–5 (1961); (k) ref. 93; (l) refs. 89 and 93; (m) ref. 54.

*RT transmission ‡77°K reflectance †80°K transmission §12°K reflectance

		E_0	e_1	E_1	E_0'	E_2	$E_2+\delta$	E_1'	
		Γ_{15}–Γ_1	L_3–L_1	Λ_3–Λ_1	Γ_{15}–Γ_{15}	X_5–X_1	X_5–X_3	L_3–L_3	d
GaP	R(a)	2·8		3·71?	4·8	5·3		6·9	23
	ϵ_2(b)	2·7		3·7?		5·2	5·5		
	CB	2·7	3·6		5·3	4·6	5·0	6·4	
GaAs	R(c)	1·35 1·75*	2·3 2·6	2·9 3·14	4·2‡	5·12‡	5·55‡	6·6 6·9	22
	ϵ_2(b)	1·35 1·65		2·8 3·1	4·2	4·7	5·1	6·3 6·6	
	CB	1·4	2·6		4·5	4·0	4·3	6·0	
GaSb	R(c)	0·70*	1·4 1·9	2·02 2·47	3·74‡	4·33‡	4·7‡	5·7	
	ϵ_2(b)	0·7	1·4 1·9	2·1 2·6	3·8	4·2	4·6	5·7	
	CB	0·8	2·3		4·4	3·8	4·2	5·6	
InP	R(d)	1·4 1·7		3·24 3·34	4·1 4·3	4·75	5·0	7·0	
	ϵ_2(b)	1·3		3·2	4·1	4·8		6·9	
	CB	1·6	2·8		4·6	4·2	4·5	6·0	
	R(e)	0·33*	2·2 2·45‡	2·48 2·76	3·9	4·92‡	5·20‡	6·4 7·0	20

	[illegible]	[illegible]		[illegible]	[illegible]	[illegible]	[illegible]	[illegible]	[illegible]
InAs	ϵ_2(b)	0·3	2·2 2·4	2·5 2·8	4·3	4·5	4·9	6·4 7·0	
	CB	0·5	2·3		4·6	3·9	4·3	5·7	
InSb	R(e)	0·18*		1·83 2·35	3·4	4·2	4·7‡	5·3 6·0	20
	ϵ_2(b)	0·2		1·9 2·5	3·4	4·0	4·5	5·4 6·0	
	CB	0·6	2·1		4·1	3·5	3·8	5·1	
ZnTe	R(f)	2·35 3·26	3·41 3·96	3·58 4·14	4·82‡	5·41		6·9 7·5	14·6
	ϵ_2(g) ‡		3·4 4·0	3·65 4·14		5·05			
	CB	2·5	4·3		6·7	5·2	5·8	7·3	
AlSb	R(h)	1·6*		2·78 3·18	3·9	4·2			
	CB	1·9	2·8		4·1	3·9	4·3	5·3	
ZnS	R(i)	3·66 3·76		5·8	7·1	7·0	7·4	9·8	13·8
	CB	3·7	5·8		8·9	6·7	7·5	9·2	
ZnSe	R(j)	2·67 3·10		4·75 5·10		6·7		9·1 9·6	13·0
	CB	2·9	5·0		7·9	6·0	6·9	8·4	
CdTe	R(f)	1·5 2·4	3·63	3·32 3·88	5·16	5·49‡	5·99‡	6·76 7·64	13·8
	CB	2·0	3·9		6·6	5·1	5·6	7·0	
HgSe	R(k)			2·90 3·21‡		6·0§	6·75§	8·3 8·6§	11·1
HgTe	R(l)		2·00 2·53‡	2·09 2·71	4·03	5·05	5·5§	7·5 8·25§	11·3
CuCl	T(m)	3·31 3·24†		6·4	6·1	8·3†			
CuBr	T(m)	2·99 3·13†		5·48 5·60†	6·3	7·3			
CuI	T(m)	3·06 3·69†		4·81 5·10†	5·98 6·40†	8·0			
AgI	T(m)†	2·92 3·76		5·00 5·40	6·80 7·71	8·7			

spin–orbit splitting of L_3' is revealed by a weak shoulder at higher energies. This transition also appears to be responsible for the structure seen near this energy in several other zincblende materials, e.g. ZnTe, CdTe, HgSe and GaP. The strong peak at 6·6 eV in HgTe (7·6 eV in HgSe) has no counterpart in the group IV and III V materials and is probably a consequence of the strong ionicity of the II VI compounds; Phillips has tentatively assigned it to transitions from the deep lying X_3 valence band into the lowest conduction band.[69]

Table 5.3 gives the interband energies for zincblende materials. The data have been taken from reflectivity data (at room temperature unless otherwise stated) for consistency. For the III V materials, energies are also given for corresponding structure

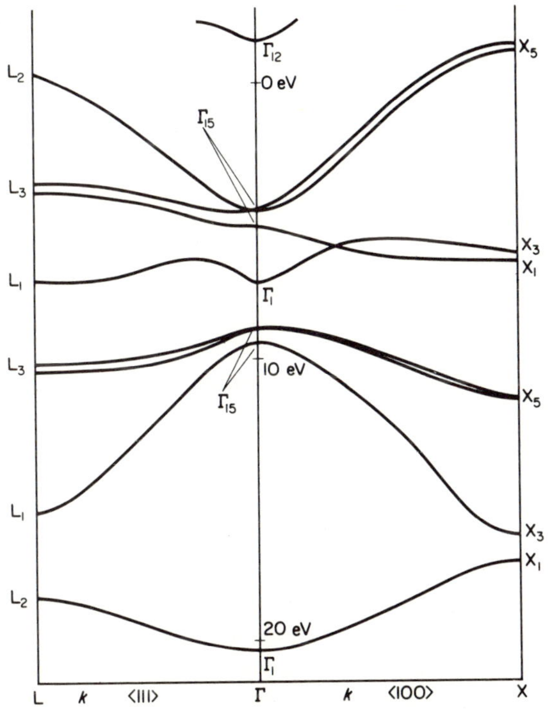

FIG. 5.6. Band structure of GaAs in the ⟨111⟩ and ⟨100⟩ directions. Note the difference in notation between the zincblende structure and the diamond structure shown in Fig. 5.3 (Cardona[101]).

in ϵ_2 (imaginary part of the dielectric constant) to illustrate the finite differences introduced by the Kramers–Kronig transform. It can be seen that for the E_2 peak and structure at higher energies the maxima in ϵ_2 mostly occur at slightly lower energies than the corresponding reflectance energies (see also ref. 95). Table 5.3 includes the theoretical estimates of the zincblende band parameters taken from the pseudopotential calculations of Cohen and Bergstresser.[(96)] In Fig. 5.6 the band structure of a typical III V semiconductor – GaAs – is shown; the difference in group theoretical nomenclature for the diamond and zincblende structures should be noted.

At higher energies excitation from levels formed by the cation *d*-electrons into the conduction band is observed and experimental values from reflectivity data for the zincblende materials are also shown in Table 5.3. In the III V compounds the *d*-electron states are separated from the lowest valence electron states by about 10 eV; in contrast, in I VII materials the *d*-electrons of the metal are heavily mixed with the halogen outer electrons and must be treated as valence electrons. This has been borne out by band structure calculations of the valence bands of AgCl and AgBr in the tight binding approximation by Bassani, Knox and Fowler[(97)] and with the APW-method by Scop.[(98)] In CuI and CuBr the mixing is even heavier so that the top valence band is probably formed mainly from the *d*-electrons of the metal rather than from the *p*-electrons of the halide ion; one should therefore be cautious in making analogies between the band structure of the I VII compounds and the other zincblende materials as different states will be involved. II VI materials represent an intermediate state and possess *d*-band excitation energies of 11–14 eV.

5.5. Systematics – isoelectronic sequences

A simple semi-empirical approach originally suggested by Herman[(99)] and discussed in more detail by Bassani and Celli[(100)] can be used to correlate the various energy gaps in the diamond and zincblende materials. The method, which predicts the behaviour of energy gaps in an isoelectronic sequence, has been used effectively to identify weak transitions from reflectivity data in a number of semiconductors.[(79,89)] The crystal potential of any group IV, III V, II VI, or I VII semiconductor can be written as:

$$V = V^{IV}_{sym} + V^{IV}_{antisym} + \lambda(V^{polar}_{sym} + V^{polar}_{antisym}). \qquad (5.1)$$

V^{IV}_{sym} and $V^{IV}_{antisym}$ are the symmetrical and antisymmetrical components (with respect to an inversion about an origin midway between the two atoms in the unit cell) of the crystal potential of the corresponding group IV element of the isoelectronic sequence being considered (e.g. Ge in the sequence Ge, GaAs, ZnSe, CuBr). $V^{IV}_{antisym}$ is zero for horizontal sequences such as the example quoted above and is zero, even for nonhorizontal sequences since the experimental results show no effect from this term. The effect of V^{polar}_{sym} is expected to be very small, consequently this term is neglected. $V^{polar}_{antisym}$ has a zero expectation value for all the states giving rise to structure in the fundamental region except those occurring at the X-point of the Brillouin zone. Thus in the zincblende materials with their lack of inversion symmetry, $V^{polar}_{antisym}$ produces the splitting of the twofold degenerate X_1 state of the diamond structure into the X_1 and X_3 states of zincblende. Cardona has also shown[101] that when the states of the double group, instead of pure orbital states, are considered, the potential for X_4 (X_5 in zincblende) states should also have a finite expectation value, corresponding to the small spin orbit splitting of X_5 in zincblende. For all other points of high symmetry in the Brillouin zone, the first-order perturbation vanishes. The perturbation coefficient λ, for the sake of simplicity, is taken equal to 1, 2 and 3, respectively, for the III V, II VI and I VII compounds. The above considerations indicate that the splitting of states at the X-point such as X_1 should be proportional to λ. Energy gaps at other points in the Brillouin zone must be obtained using second-order perturbation theory. Let us consider two states which interact through a purely polar antisymmetric potential, i.e. states belonging to different irreducible representations in diamond but belonging to the same irreducible representation in zincblende. Whereas the matrix element of the potential between the wave functions in the diamond structure is zero the corresponding matrix element in zincblende is different from zero. Examples of this are Γ'_{25}–Γ_{15} (Γ_{15}–Γ_{15} in zincblende) and

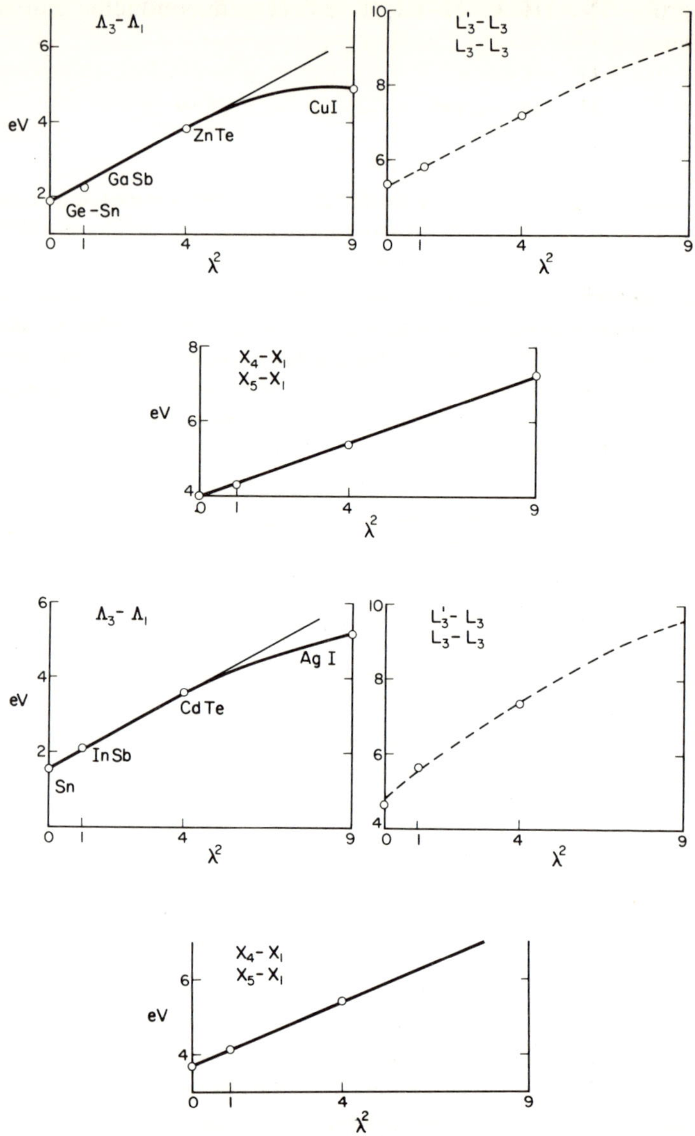

FIG. 5.7. Systematic variation of energy gaps as a function of λ^2 for the isoelectronic sequences GeSn–GaSb–ZnTe–CuI and Sn–InSb–CdTe–AgI (Cardona and Greenaway[89]). (○ = experimental points,-----calculated curve.)

L_3'–L_3 (L_3–L_3 in zincblende). If one regards only the mutual repulsion of the two states concerned (i.e. it is assumed that the states are well separated from other states of the same symmetry) then by solving a 2×2 secular equation we obtain the following expression for the polar energy gaps:

$$E_p = E_{np}\left[1+\frac{4\lambda^2|\langle V_p\rangle|^2}{E_{np}{}^2}\right]^{1/2}, \tag{5.2}$$

where E_{np} is the energy gap of the unperturbed group IV member of the isoelectronic sequence and $\lambda^2|\langle V_p\rangle|^2$ the square of the matrix element of the perturbing polar potential between the two states involved in the transition. Equation (5.2) reduces for small perturbations, i.e. for a polar potential small compared with the gap between the interacting states, to

$$E_p = E_{np}+\frac{2\lambda^2|\langle V_p\rangle|^2}{E_{np}}. \tag{5.3}$$

For an energy gap between states which interact neither in diamond nor in zincblende there is clearly no direct influence of the polar potential. Considering only the orbital interaction, Λ_3–Λ_1 and X_5–X_1 are examples of this situation. The spin–orbit interaction however causes a small mixing in both these two cases, so that one is nevertheless justified in taking equation (5.3) for the gaps, provided that other states of the same symmetry are well separated. The effectiveness of this semi-empirical approach can be seen in Figs. 5.7 and 5.8 where the energies of three of the main interband transitions ($\Lambda_3-\Lambda_1$, X_4-X_1, $L_3'-L_3$) have been plotted as a function of λ^2 for three isoelectronic sequences in the periodic table. The solid lines represent the behaviour expected from equations (5.2) and (5.3) and the open circles the experimental values from reflectivity data. We see that the treatment holds not only for normal sequences but also for skew sequences such as (α-Sn)–Ge, GaSb, ZnTe, CuI, where the energy gap of the group IV member is taken as the mean value of the gaps of the two group IV components. The expected proportionality of the X_1–X_3 splitting to λ for the sequence Ge, GaAs, ZnSe, CuBr is also shown in Fig. 5.8. We see also that when applicable, equation (5.3) holds up to a point where the perturbation is about equal in

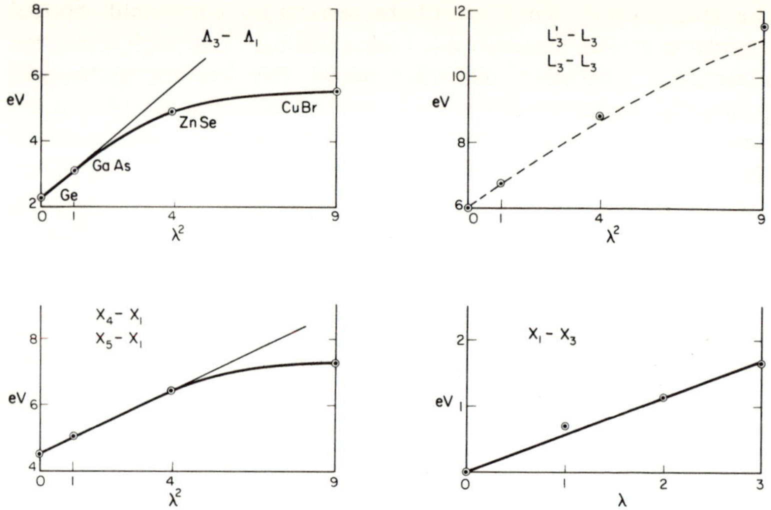

FIG. 5.8. Systematic variation of energy gaps as a function of λ^2 and λ for the isoelectronic sequence Ge–GaAs–ZnSe–CuBr. (○ = experimental points, ----calculated curve.)

magnitude to the non-polar gap. These results also indicate that the band structure of the I VII materials is similar to that of the other zincblende semiconductors, even though the top valence bands are formed from cation *d*–bands in the case of the former, and from anion *p*-bands in the latter materials. In this simple approach λ is given the integral values 1, 2, and 3, corresponding to equal increases in polarity or ionicity as one moves further away from the group IV material. Phillips has pointed out[69] that the success of this method depends on the cancellation of effects outside the simple two band model and in fact λ is not expected to have exactly integral values.

5.6. Group IV VI and group V materials

5.6.1. *Lead salts* PbS, PbSe *and* PbTe

The lead salts are polar semiconductors, and with ten electrons in the unit cell, crystallize in the rocksalt structure (two interpenetrating face-centred cubic lattices with the origin of the chalcogen lattice located at the midpoint of the body diagonal of the lead lattice). They are characterized by very small optical energy gaps (see Table 5.4.), high mobilities and high dielectric constants. Extensive investigation of the low-energy optical properties and transport properties[102] have established that the absolute extrema of both valence and conduction bands occur at the L point of the Brillouin zone for all three lead salts. The direct optical gaps of all three compounds possess an unusual positive temperature coefficient in contrast to the negative coefficient observed for diamond and zincblende materials, and this is a good

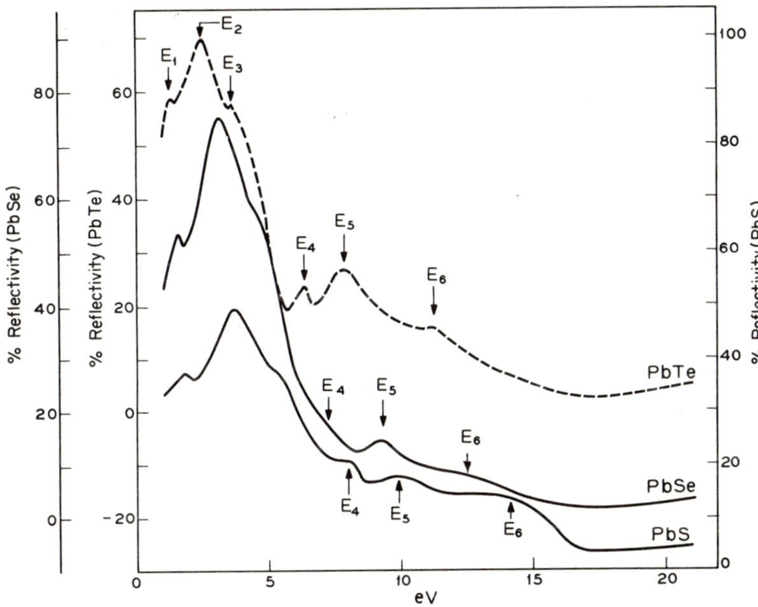

FIG. 5.9. Reflectivity of the lead salts PbS, PbSe and PbTe at 300°K (Cardona and Greenaway[10]).

indication that overall similarities exist in the band structure of the lead salts. The close similarity was realized in early thin film measurements of absorption[103] and more recently in the fundamental reflectivity spectra.[10] The room temperature reflectivity is shown in Fig. 5.9 where the maxima have been labelled E_1–E_6; in all cases E_2 corresponds to the absolute maximum in reflectivity. A convenient way of illustrating the reflectivity data is shown in Fig. 5.10, where the E_1–E_6 maxima and also E_0, the fundamental absorption edge, are plotted as a function of lattice parameter. Figure 5.10 also includes the results for the allied materials SnTe and GeTe which are discussed in the following section. It is quite clear that corresponding transitions have the same origin in the lead salts; the only anomaly is the optical gap, which is lower in PbSe than in the other two materials. It is tempting to interpret the reflectivity data by analogy with the zincblende materials, and an attempt in this direction[10] was made when the results of a first-principles augmented plane wave (APW) band calculation

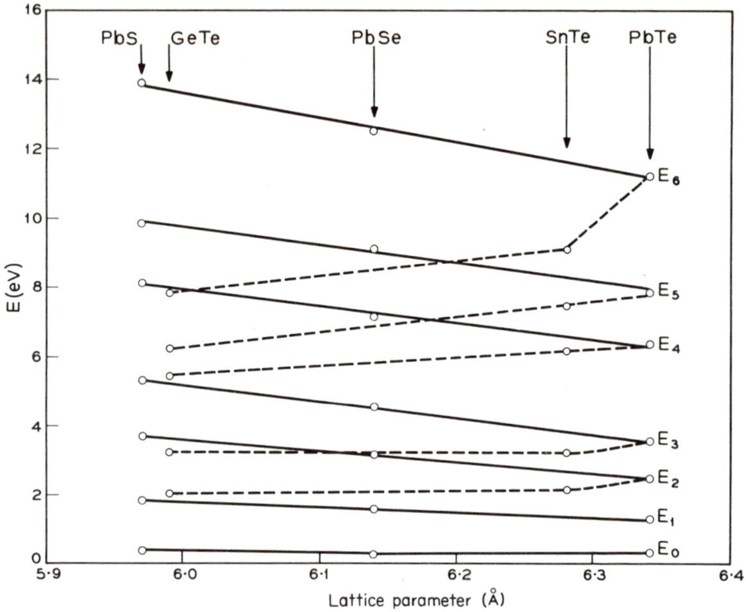

FIG. 5.10. Variation of energy gaps with lattice parameter for the three lead salts, GeTe and SnTe (Cardona and Greenaway[10]).

for PbTe by Johnson *et al.*[104–5] became available. Unfortunately one cannot make direct analogies with the zincblende materials as the band ordering is quite different in the lead salts. The *s*-band which was the lowest conduction band in zincblende falls below the top *p*-like valence band in the lead salts because of the low energy of the s^2 subshell in heavy metal atoms. From the standpoint of a basic calculation, it is extremely difficult to obtain crystal potentials for such heavy atoms with an accuracy commensurate with the small band gap of PbTe. The difficulties are increased by the large spin–orbit interaction in PbTe and also by the presence of other large relativistic effects which, as Johnson *et al.* realized, can cause uncertainties in the ordering of states and the consequent location of the Fermi level. One important feature of the band structure—the spin–orbit splitting at the zone centre—was, however, identified with the E_4 and E_5 maxima of Fig. 5.9. These peaks were assigned to transitions from a deep Γ_1 valence band formed from the $5s^2$ tellurium electrons, into a spin–orbit

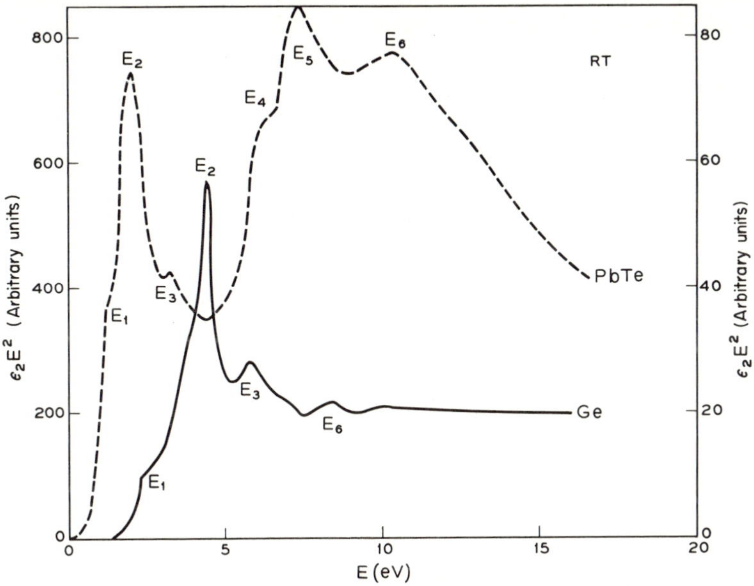

FIG. 5.11. The density of states function $\epsilon_2 E^2$ for Ge and PbTe derived from normal incidence reflectance data.

split Γ_{15} conduction band formed mainly from $6p^2$ lead electrons. Two pieces of evidence supported this assignment: In Fig. 5.11 the density of states function $\epsilon_2 E^2$ is plotted against photon energy E for PbTe and Ge (it is assumed that the matrix elements of the strong interband transitions are approximately independent of E). It is seen that the E_4 and E_5 maxima in PbTe constitute two extra peaks in otherwise broadly similar curves, and one can logically assign them to transitions involving the two $6s$ electrons from Pb in the PbTe unit cell which according to ref. 104 form a deep-lying Γ_1 valence band. Secondly, as the spin–orbit split Γ_{15} conduction band is formed mainly from 6*p* lead electrons, one might expect the separation between the E_4 and E_5 maxima to remain constant in the three lead salts, but to decrease for the structurally allied, but lighter compounds SnTe and GeTe. Figure 5.10 clearly shows that this behaviour takes place. Theoretical confirmation of this assignment has been obtained by Lin and Kleinman [102] in a pseudopotential band calculation for all three lead salts which included all the important relativistic effects and the spin orbit interaction, and involved only five adjustable band parameters (including a spin–orbit parameter). Using the above value for the spin–orbit splitting Lin and Kleinman were able to predict the trends as one goes from PbS to PbTe, locate the main interband

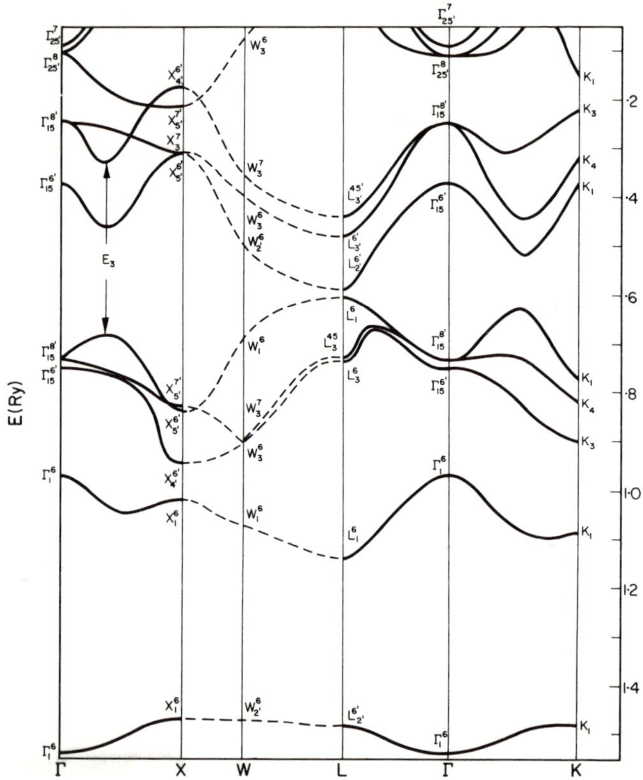

FIG. 5.12. Band structure of PbS calculated by Lin and Kleinman.[102] Both the single group (subscript) and double group (superscript) notations of the energy states are given.

transitions, and obtain extremely good agreement with the reflectivity data. The comparison between the theoretical and experimental results is given in Table 5.4. The calculation is also able to explain the anomaly between the optical gaps of the three lead salts: the lowest conduction band at L is formed from L_2' states for PbS and PbSe but from L_3' states in PbTe. The pseudopotential calculation results for PbS over the Brillouin zone are shown in Fig. 5.12. The band structures for PbSe and PbTe are qualitatively similar except that the two lowest conduction band states L_2' and L_3' are inverted in PbTe.

TABLE 5.4

Energy gaps in the lead salts PbS, PbSe and PbTe, experimental and calculated. The E_1–E_6 experimental results are from room temperature reflectance data and the theoretical values from the Lin and Kleinman[102] *calculation. E_0 values are taken from room temperature absorption data. Both the single group (subscripts) and double group (superscripts) notations of Lin and Kleinman are given*

	Transition	Experimental value (eV)			Calculated value (eV)		
		PbS	PbSe	PbTe	PbS	PbSe	PbTe
E_0	$L_1^6(2)$–$L_*^{6'}$	0·37	0·26	0·29	0·37	0·26	0·29
E_1	$L_1^6(2)$–$L_{**}^{6'}$	1·83	1·54	1·24	1·77	1·52	1·31
E_2	$L_3^6(1)$–$L_{**}^{6'}$	3·67	3·12	2·45	3·5	3·34	2·52
E_3	$\Delta_1^6(3)$–$\Delta_5^6(2)$	5·3	4·5	3·5	5·3	4·6	3·59
E_4	$\Gamma_1^6(1)$–$\Gamma_{15}^6(2)$	8·1	7·1	6·3	8·15	7·3	6·3
E_5	$\Gamma_1^6(1)$–$\Gamma_{15}^8(2)$	9·8	9·1	7·8	9·85	8·96	7·8
E_6	$X_{5'}^{7'}(1)$–$X_2^7(1)$	13·9	12·5	11·2	13·9	13·2	11·4

$L_*^{6'} = L_{3'}^{6'}(1)$ in PbTe and $L_{2'}^{6'}(2)$ in PbS and PbSe.
$L_{**}^{6'} = L_{2'}^{6'}(2)$ in PbTe and $L_{3'}^{6'}(1)$ in PbS and PbSe.

5.6.2. SnTe, GeTe *and group V semimetals* As, Sb *and* Bi

SnTe crystallizes in the rocksalt structure but in GeTe a small trigonal distortion of about 1·5° is present. The group V semimetals As, Sb and Bi also have the quasi-rocksalt structure, but with a trigonal distortion of about 3°. Although these materials are noncubic (with the exception of SnTe) it is convenient to treat them with the lead salts, as marked similarities in the overall band structure exist.[10,106] These similarities are clearly illustrated in Fig. 5.13 where the fundamental reflectivity curves of

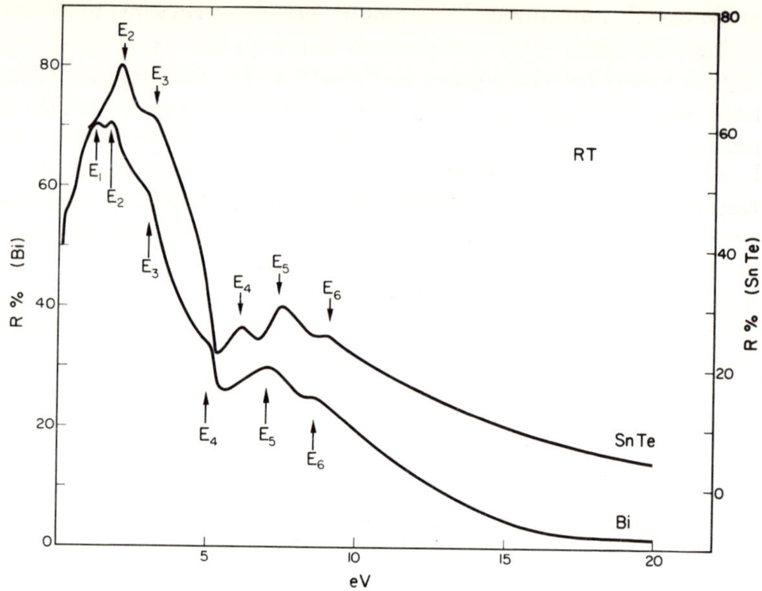

FIG. 5.13. Reflectivity spectra of cleaved Bi and SnTe at 297°K (Cardona and Greenaway[(10)]).

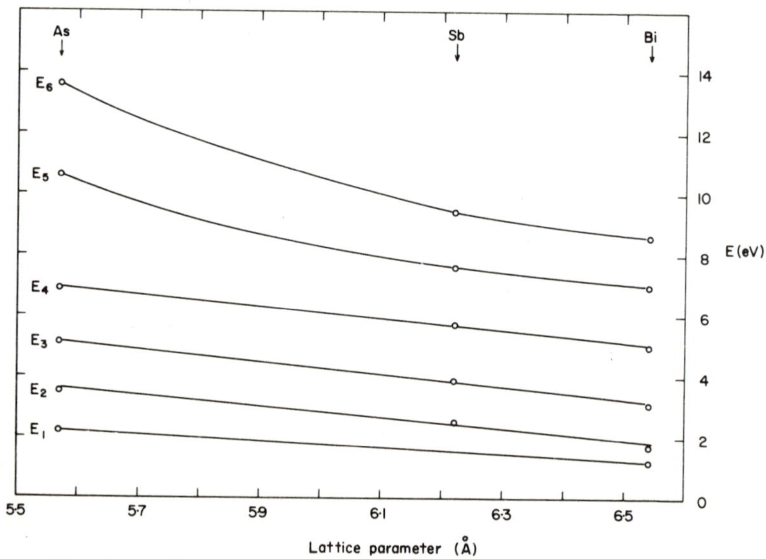

FIG. 5.14. Variation of energy gaps with lattice parameter for As, Sb and Bi (Cardona and Greenaway[(10)]).

SnTe and Bi are shown. The maxima have been labelled using the lead salt notation. The systematic behaviour of the peaks is shown in Fig. 5.10 which compares GeTe and SnTe with the lead salts, and Fig. 5.14 which shows the peak energies as a function of lattice parameter for As, Sb and Bi. From these results it is likely that the band structure derived for the lead salts also applies to these materials, and the effect of the anisotropy on the band structure is minimal. First pseudopotential band structure calculations have been performed by Falicov and Lin and by Lin and Phillips,[107] and indeed show that the energy bands in the group V and IV VI materials are similar. For crystalline antimony, excellent agreement is obtained between reflectivity data and the pseudopotential results. It is seen from Fig 5.10 that when the lead atom is replaced by the lighter Sn and Ge, all energy gaps decrease. This striking behaviour is contrary to the results for diamond and zincblende materials where the lowest conduction band has *s* character and the opposite trend is observed. The large effect in the E_4 and E_5 maxima, for example, is associated with the presence of an *s*-like valence band and a predominantly *p*-like conduction band. With increasing atomic number, the energy of the *s*-states decreases (moves nearer to the core) much faster than the *p*-states, consequently energy gaps increase in the series GeTe–SnTe–PbTe. Comparison between the reflectivity of the lead salts (Fig. 5.4.), SnTe and GeTe and the group V semimetals, shows a rise in reflectivity above 15 eV for the lead salts which is absent in the other materials. In the lead salts the effect is due to transitions from filled atomic *d*-bands into the conduction band. These *d*-bands are located about 24 eV below the conduction band, and the effect is completely analogous to the behaviour observed in diamond and zincblende semiconductors and discussed in the previous sections. In SnTe and GeTe the *d*-bands lie deeper and should not be observed over the energy range studied here: the reflectivity results support this. Again, in As, Sb and Bi no *d*-band effects are visible, as the *d*-bands lie $\sim$ 50 eV below the conduction band.

5.7. Alkali halides

The alkali halides are characterized by large optical gaps and low refractive indices. The Li, Na, K and Rb halides all have the rocksalt structure; the caesium halides are also cubic but possess the simple cubic crystal structure. Because of the larger effective masses and low dielectric constants, one expects the absorption of photons to produce large Coulomb interactions between holes and electrons. Thus conventional excitons play a dominant role in the optical spectra of the alkali halides, and it is these exciton effects which have been recognized for many years,[108] and cause a large fraction of the fundamental absorption. Recently new types of exciton behaviour have been proposed to explain certain features of the fundamental optical spectra both in the alkali halides and in the diamond and zincblende semiconductors. We have, therefore, divided the discussion of the alkali halides into direct interband scattering spectra and interband exciton effects which are degenerate with the main scattering continuum. In this section we deal only with the former; exciton effects are treated in Chapter 8.

Thin single crystal films of the majority of the alkali halides can easily be prepared by evaporation techniques, consequently by using LiF substrates, absorption results of good accuracy can be obtained out to about 12 eV. No suitable substrates exist for measurements above the LiF absorption edge, thus at high energies reflectivity data afford the most practical way of investigating the band structure. Comprehensive absorption data at room temperature and 80°K for all alkali halides (except LiF) have been obtained by Eby, Teegarden and Dutton[46] who used an elegant differential measurement technique to eliminate possible effects of surface contamination. Room temperature reflectivity spectra for KCl, KBr and KI were obtained by Philipp and Ehrenreich[109] out to an energy of 25 eV. As with the diamond and zincblende semiconductors, the rocksalt structure alkali halides reveal marked similarities in their optical spectra and hence band structure. The fundamental absorption and reflectivity spectra of the majority of the halides are dominated by a sharp low-energy exciton doublet caused by *p*-valence states associated with the halogen ions. The splitting is due to spin–orbit interaction and closely follows the atomic splitting; it can be identified with the threefold degenerate top Γ_{15} valence band at the centre of the Brillouin zone (i.e. identical to the situation in diamond and zincblende materials). The presence of this characteristic splitting gives an immediate way of identifying further transitions involving the top valence band either at the zone centre or in the $\langle 111 \rangle$

direction. Again, the splitting at the L point should be two-thirds the splitting at the zone centre. For most of the alkali halides the direct interband threshold $\Gamma_{15}(\frac{3}{2})-\Gamma_1$ associated with the first large exciton peak has been clearly identified as a weak M_0 shoulder in the low-temperature absorption spectra. For many materials the spin–orbit split counterpart $\Gamma_{15}(\frac{1}{2})-\Gamma_1$ is also seen

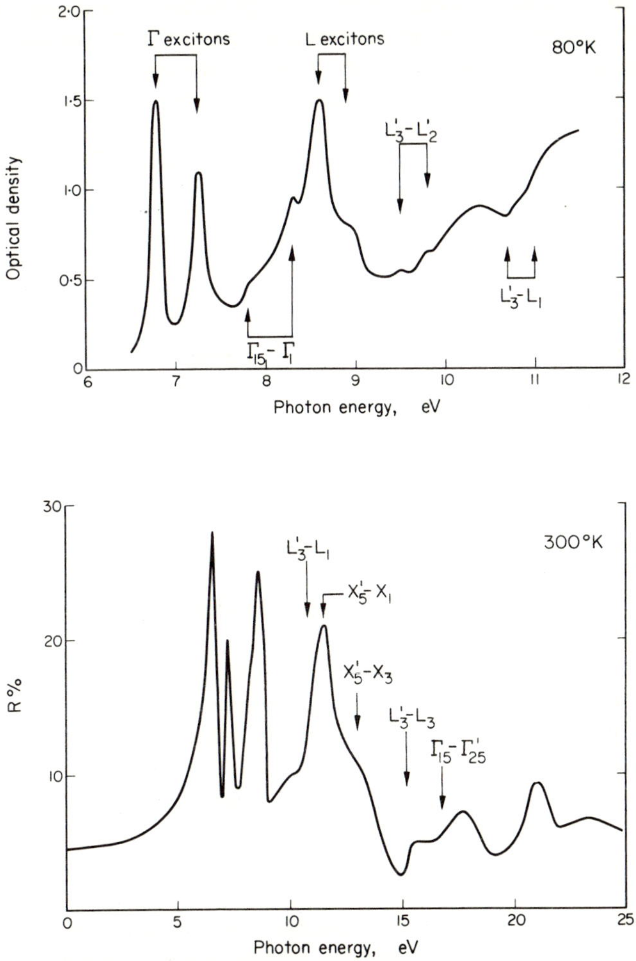

FIG. 5.15. Absorption and reflectance spectra of KBr. The 80°K data were obtained by Eby, Teegarden, and Dutton[46]; the 300°K data by Philipp and Ehrenreich.[109]

at higher energies. The 80°K absorption curve for KBr is shown in Fig. 5.15. The two thresholds at 7·8 eV and 8·3 eV are well separated from the adjoining stronger transitions. Figure 5.15 also shows the room-temperature reflectivity of KBr out to an energy of 25 eV. The assignments given for the structure above the direct thresholds are the result of a detailed analysis by Phillips[69,110] of the spectra of the whole family of rocksalt structure alkali halides. These assignments clearly reveal the close similarity with diamond and zincblende materials. In particular the dominant high-energy peak in the series KCl, KBr and KI is due as in the diamond structure to X_5'–X_1 transitions

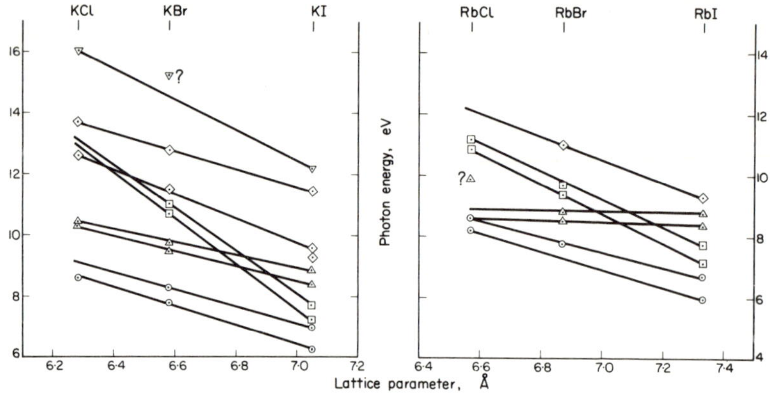

⊙ = Γ_{15}–Γ_1, ⊡ = L_3'–L_1, △ = L_3'–L_2', ◈ = X_5'–X_1X_3, ▽ = L_3'–L_3

FIG. 5.16. Variation of energy gaps with lattice parameter for the sequences KCl, KBr, KI, and RbCl, RbBr, RbI.

(X_5–X_1 in zincblende). The insensitivity of the magnitude and line shape of this peak in all the group IV, III V and II VI materials and the rocksalt alkali halides is almost certainly due to the face-centred cubic lattice truncated octahedral Brillouin zone which is shared by the whole series of materials. In the alkali halides one also expects to observe the splitting of the conduction band at X into X_1 and X_3 states, and this is clearly seen as a high-energy shoulder at about 13·3 eV on the 11·5 eV peak of Fig. 5.15. In the study of the lead salts and allied compounds, it is instructive to display the energies of the interband transitions as a function of lattice parameter; this technique can also be used for the alkali halides, and forms a valuable method of identifying some of the weaker transitions. Figure 5.16 shows the results for most of the interband transitions in the two sequences KCl, KBr, KI, and RbCl, RbBr, RbI. (The two fluorides KF and RbF have much smaller lattice parameters – 5·33 Å and 5·63 Å respectively – and insufficient experimental data is available to include them in the sequences, but values of the lowest $\Gamma_{15}(\frac{3}{2})$–Γ_1 threshold obtained from ref. 46 fit an extrapolation of Fig. 5.16 to within a few tenths of an eV.) These elementary systematics provide further illustration of the close identity between the alkali halides and enable estimates to be made of the positions of transitions which are obscured in the optical spectra either by virtue of low oscillator strength or degeneracy with stronger transitions.

For KI, both reflectivity and absorption data reveal the expected small X_5' valence band splitting due to spin–orbit interaction. This splitting is presumably too small to be resolved in KBr and KCl. The systematics for the rubidium halides indicate

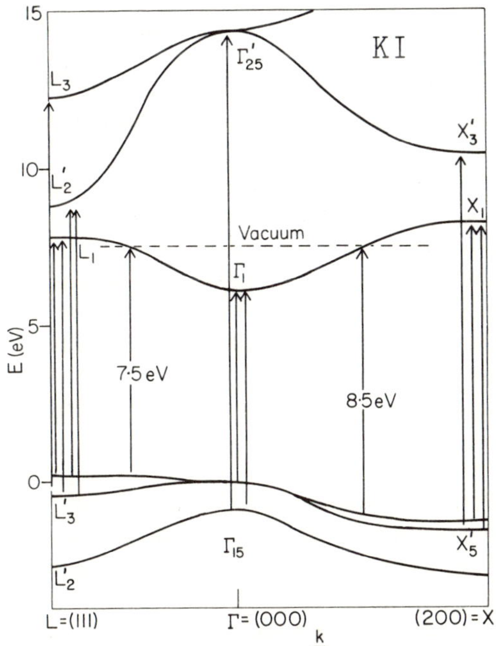

FIG. 5.17. Band structure of KI in the ⟨111⟩ and ⟨100⟩ directions (Phillips[110]).

that the lowest $\Gamma_{15}(\frac{3}{2})$–Γ_1 threshold for RbBr is near 7·3 eV and the value given in ref. 110 corresponds to the $\Gamma_{15}(\frac{1}{2})$ transition. The shape of the energy bands along the $\langle 111 \rangle$ and $\langle 100 \rangle$ axes has been obtained for the potassium halides by Phillips[(110)] using a semiempirical approach which regarded the valence bands as a narrower version of the bands in zincblende crystals, and by drawing parallels with the band structure of the solid rare gases.[(111)] The energy bands for KI are shown in Fig. 5.17. Similar schemes apply for KBr and KCl with the exception that for the lowest conduction bands at L, L_1 and L_2' are interchanged. These band models indicate that the absolute maximum of the valence band lies near the L-point. One would thus expect weak indirect absorption thresholds about 0·3 eV below the direct thresholds. These thresholds are obscured experimentally by the strong background caused by the first exciton peak.

It should be noted that in Phillips' discussion the conduction bands are assumed to be free-electron like. Onadera *et al.*[(112)] have recently calculated the band structure of KI using a relativistic Green's function method. They emphasize that the d-conduction bands originating from the $3d$ states of potassium lie just above the s-like conduction bands. This leads the authors to a different assignment for the second exciton doublet, namely to excitons associated with M_0 edges at X instead of M_1 edges at L. The fact that the corresponding exciton lines are found in other potassium and rubidium halides but not in sodium halides, where no such low-lying d levels exist, lends support to this assignment, but at the present time it is not possible to decide unequivocably whether this more recent model is correct. For this reason the original assignments due to Phillips are shown in Figs. 5.15–5.17 (and also in Fig. 8.4).

The collected values for the interband transitions in the rock-salt alkali halides are shown in Table 5.5. These values have been taken from both reflectivity and absorption data; the bracketed gaps are those which have been obtained by inference or extrapolation from lattice parameter versus energy gap plots. No attempt has been made to extrapolate the sequences to the alkali fluorides.

The caesium halides (with the exception of CsF) crystallize in the simple cubic structure. The low-temperature absorption spectra reveal the sharp exciton peaks associated with the spin–orbit

TABLE 5.5

Interband transitions in the rocksalt structure alkali halides (according to Phillips[110]) taken from 80° K absorption data for $E < 11$ eV and from 300°K reflectivity data for $E > 11$ eV

	NaCl	NaBr	NaI	KF	KCl	KBr	KI	RbF	RbCl	RbBr	RbI
$\Gamma_{15}(\frac{3}{2})-\Gamma_1$	8·4	7·5	6·75	10·9	8·6	7·8	6·3	10·5	8·2	(7·3)	6·0
$\Gamma_{15}(\frac{1}{2})-\Gamma_1$		7·9	7·0	(11·3)	(9·1)	8·3	7·0		8·6	7·8	6·75
$L_3'(\frac{3}{2})-L_1$			8·25		(12·9)	10·7	7·25		10·9	9·4	7·2
$L_3'(\frac{1}{2})-L_1$			9·15		(13·1)	11·0	7·75		11·2	9·7	7·8
$L_3'(\frac{3}{2})-L_2'$	10·3				10·35	9·5	8·45		(9·9) ?	8·55	8·4
$L_3'(\frac{1}{2})-L_2'$					10·45	9·8	8·9			8·85	8·8
$L_3'-L_3$					16·0	15·2	12·2				
$X_5'-X_1$					12·6	11·5	9·3, 9·6		(12·1)	11·0	9·3
$X_5'-X_3$					13·7	12·8	11·5				
$\Gamma_{15}(\frac{3}{2})-\Gamma_{25}'$						17·0	14·3				
$\Gamma_{15}(\frac{1}{2})-\Gamma_{25}'$							15·2				

split Γ_{15}–Γ_1 transitions but the splittings are anomalous owing to configuration interactions and do not correlate simply with the free halogen splittings. No complete band calculations are yet available for the caesium halides but Phillips has made an initial analysis of the valence and conduction band structure on the basis of results for the face centred cubic alkali halides and solid rare gases.[113] The absorption spectrum of CsBr is shown in Fig. 5.18 and Phillips' assignments for the low-energy portion are given; the origin of the structure above about 9 eV is obscure, but a number of van Hove singularities can be recognized. The direct

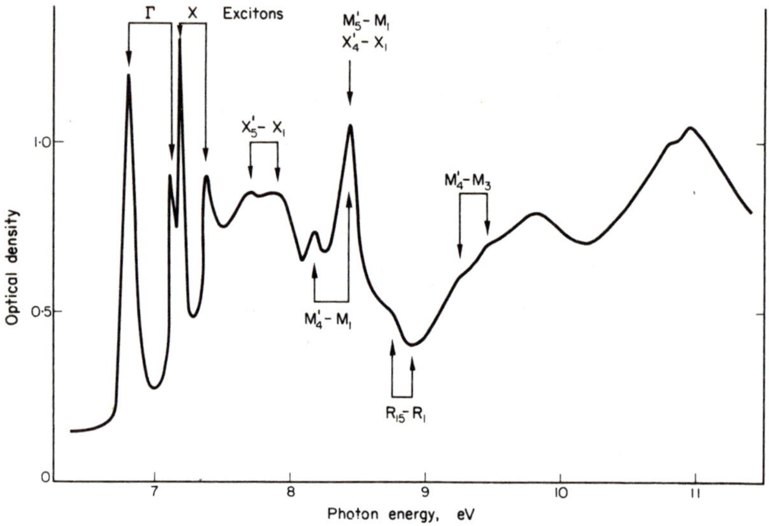

FIG. 5.18. 80°K absorption spectrum of CsBr (Eby, Teegarden and Dutton[46]).

Γ_{15}–Γ_1 absorption thresholds are masked by the strong Γ exciton peaks. A similar spectrum is obtained for CsCl and the interband transitions can be assigned by analogy with the CsBr data. CsI possesses a rather different absorption spectrum, probably caused by the lowering of the *d*-conduction band levels by the iodide ions; however, all the low-energy transitions seen in CsBr (and also one of the Γ_{15}–Γ_1 thresholds) can be recognized.

CHAPTER 6

FUNDAMENTAL OPTICAL SPECTRA OF ANISOTROPIC SEMICONDUCTORS AND INSULATORS

6.1. General

Crystals of hexagonal, tetragonal and trigonal crystal classes have one optical axis; crystals of all other non-cubic classes are optically biaxial. The optical properties of the anisotropic crystals which now also include birefringence and dichroism are described by the tensorial complex dielectric function $\epsilon = \epsilon_1 + i\epsilon_2$. The tensors ϵ_1 and ϵ_2 can always be brought into a diagonal form, and for crystals of rhombic or higher symmetry the main axes of the tensors ϵ_1 and ϵ_2 coincide.

Formulae for the reflectivity equivalent to equations (2.15) and (2.16) can only be given in simple cases, and a determination of the wavelength dependent tensor N from measurements under oblique incidence has not been achieved so far even for uniaxial crystals. It can, however, be done relatively simply from the normal incidence reflectivity of linearly polarized light where one orients the electric vector of the light parallel to the different crystallographic axes. Three spectra result for biaxial crystals and two for uniaxial crystals since two crystallographic axes are equivalent. The evaluation of the tensors n and k or ϵ_1 and ϵ_2 is then straightforward using the Kramers–Kronig relations. The interpretation of the ϵ_2 spectra proceeds along the same lines as for cubic materials and since for a given crystal, the band structure and thus the joint density of states is given, the differences between the components of the tensor ϵ_2 arise only from differences in the matrix elements governing the transitions. As outlined in Section 4.6, in most cases the wave functions in the matrix element M_{vc} in equation (4.22) are not explicitly known, but we can determine from the character tables whether a given transition is allowed or forbidden for a certain orientation of the electric vector according to equation (4.23). We find that the selection rules thus provide a further parameter for the evaluation of the band structure. One way to use this parameter is to study anisotropic crystals which can be regarded as slightly distorted cubic systems. We can then use the isoelectronic sequence scheme discussed in Section 5.5 and treat the anisotropy effects within a perturbational framework. In principle the selection rules reduce the number of possible assignments of interband transitions to structure in the optical spectra in every anisotropic crystal. In crystals of low symmetry the selection rules may even be used to determine some aspects of the band structure without existing feedback from a pertinent band structure calculation.

6.2. Uniaxial crystals

6.2.1. *Wurtzite materials*

Some of the II VI compounds, e.g. ZnS, CdS and CdSe, crystallize in the hexagonal wurtzite structure (point group C_{6v}). The actual atomic distances are such that one can also regard the structure as a cubic zincblende lattice slightly deformed along the body diagonal. Consequently the long-wavelength birefringence is only about 0·5%[114] and the energy separation for both absorption edges under the two orientations of the electric vector with respect to the c-axis is only a few hundredths of an eV.[115–7] The exciton structure associated with the band edges has been extensively studied[116–7] and was reviewed by Reynolds *et al.*[118] A strong similarity between the band structure of zincblende and wurtzite crystals can also be expected from theoretical grounds. Birman[119] suggested that the band structure of the wurtzite crystals at points in **k**-space along the hexagonal axis can be derived by perturbation theory from that of the zincblende type crystals along a $\langle 111 \rangle$ direction. Hence one expects the electronic states to be very similar at $\mathbf{k} = 0$ and along the two corresponding directions.

The fundamental reflectivity spectra of hexagonal CdS both for $E \parallel c$ and $E \perp c$ are shown in Fig. 6.1.[120] There is an overall similarity with the spectrum of cubic CdS[121] which confirms the viewpoint taken by Birman. The assignments of the reflectivity peaks E_0, E_0', E_1, E_1' and E_2 can therefore be based on the ana-

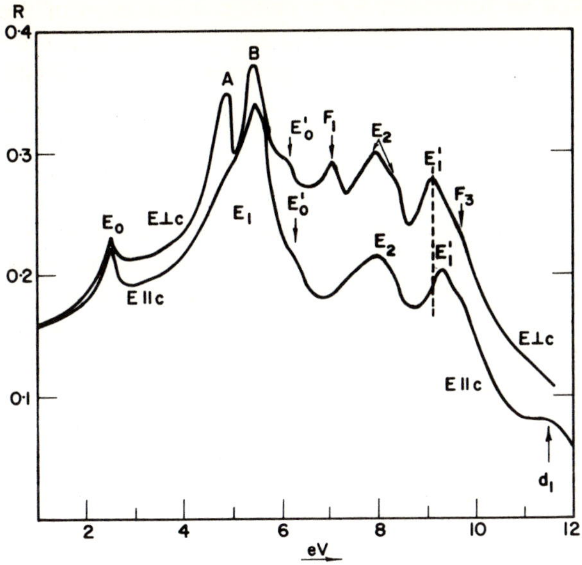

FIG. 6.1. Room temperature reflectivity of hexagonal CdS for $E \perp c$ and $E \parallel c$ (Cardona and Harbeke[(120)]).

lysis of the optical spectra of zincblende materials using the isoelectronic sequence scheme outlined in Section 5.5, and the notation of the peaks is therefore the same as for the cubic II VI compounds. The subscripts 0, 1 and 2 of E designate transitions occurring at the centre of the zone Γ, in the $\langle 111 \rangle$ direction, and the $\langle 100 \rangle$ direction of $\mathbf{k}$ space, respectively.

There are, however, also strong polarization effects: the E_1 peak of cubic CdS is split into two components A and B for $E \perp c$ and the A component is not present for $E \parallel c$. In CdSe even the expected spin–orbit splitting of the Λ_3 band is seen in the experiment.[(122)] The E_1' peaks in the two spectra are slightly shifted with respect to each other, the energy of the peak for $E \parallel c$ being 0·2 eV higher than for $E \perp c$. Finally, the F_1 peak only appears for $E \perp c$.

In Table 6.1 the energies of the various peaks in R and k (or ϵ_2) from refs. 120 and 121 are compiled. Normal incidence reflection spectra have also been taken with unpolarized light[(122)]; the results with the light propagating parallel to the c-axis agree well with the data in Table 6.1.

TABLE 6.1

Energies of the peaks in extinction coefficient k (or ϵ_2) and reflectivity R observed at room temperature in ZnS, CdS *and* CdSe (in eV)

		ZnS[a]			CdS			CdSe[a]	
		Cubic	Hexagonal		Cubic[b]	Hexagonal[a]		Hexagonal	
			$E\|\|c$	$E\perp c$		$E\|\|c$	$E\perp c$	$E\|\|c$	$E\perp c$
		3·66	3·74	3·78					
E_0	R	3·76	3·88	3·87	2·50	2·53	2·53	1·9	1·85
E_0'	k				6·4(ϵ_2)	6·2	6·1	6	
E_0'	R				6·2	6·3	6·2	6	
E_1 A	k	5·82					4·98		4·31, 4·08
E_1 A	R	5·79					4·93		4·30, 4·07
E_1 B	k				5·3(ϵ_2)	5·48	5·50	4·98	4·85
E_1 B	R		5·6		5·49	5·52	5·50	5·00	4·82
E_1'	k	9·65	9·56	9·43	8·8(ϵ_2)	9·27	9·00	8·75	8·50
E_1'	R	9·78	9·73	9·61	9·18	9·35	9·15	8·63	8·35

TABLE 6.1 (*Continued*)

		ZnS[a]			CdS			CdSe[a]	
		Cubic	Hexagonal		Cubic[b]	Hexagonal[a]		Hexagonal	
			$E\|\|c$	$E\perp c$		$E\|\|c$	$E\perp c$	$E\|\|c$	$E\perp c$
E_2	k	7·03, 7·35	7·01, 7·50	7·00, 7·5	7·5 (ϵ_2)	7·8	7·8	7·30	7·54
	R	6·99, 7·41	6·98, 7·56	7·00, 7·52	7·95	8·04, 8·35	8·00, 8·35	7·50	7·55
d_1	k	10·6			11·6(ϵ_2)	11·4			
	R	10·8	10·8	10·8	11·9	11·5		10·75	10·6
d_2	k	14·3			13·3(ϵ_2)				
	R	13·8	13·8	13·8	14·4	14·0	14·0	14·0	14·0
F_1	k			6·6					6·9
	R			6·6			7·12		6·8
F_2	k	7·9							
	R	7·9			9·6 (ϵ)				
F_3	R				9·7	9·8	9·8	9·35	9·2

[a] Ref. 120. [b] Ref. 121.

To discuss the polarization effects which are most pronounced at the peaks E_1 and E_1', i.e. for transitions occurring at points in the $\langle 111 \rangle$ direction, it is instructive to compare the Brillouin zones of both crystal types. Figure 6.2 shows the truncated octahedron of face-centred cubic crystals and the hexagonal prism of wurtzite. The wurtzite unit cell contains twice the number of atoms of the zincblende cell and hence the Brillouin zone has about half the volume of that of zincblende. Following the double-zone scheme of Birman[(119)] we therefore obtain an approximate fit by putting two wurtzite zones together in the $\langle 001 \rangle$ direction. There is then a correspondence between the vector ΓL in zincblende and the vector $\Gamma A \Gamma'$ in wurtzite. In the reduced wurtzite zone Γ' coincides with Γ and hence the point L in zincblende is equivalent to Γ in wurtzite, i.e. the distance $A\Gamma'$ is folded into $A\Gamma$. The number of Γ states is thus double that of zincblende. Half of the Γ states correspond directly to Γ states of zincblende. The triplet Γ_{15} states of zincblende correspond to a doublet Γ_5 and a singlet Γ_1. The Γ_5–Γ_1 splitting of the top valence band is obtained from the splitting of the absorption edge. The other half of the states correspond to L states of zincblende. These have, respectively, Γ_6 (from L_3) and Γ_3 (from L_1) symmetry. Such correspondence can be found by comparing the irreducible representations of the transformations which are common to both groups.[(60)] In

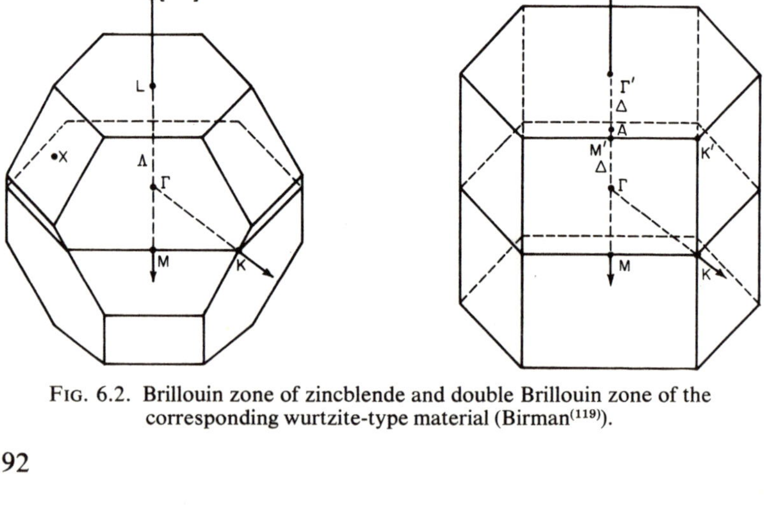

FIG. 6.2. Brillouin zone of zincblende and double Brillouin zone of the corresponding wurtzite-type material (Birman[(119)]).

the same way the Λ_3 and Λ_1 states involved in the transition causing the E_1 peak in zincblende correspond to Δ_5 and Δ_1 or to Δ_6 and Δ_3 states, respectively, depending on which side of A they are located in the double zone. If we now calculate the selection rules for Δ_5–Δ_1 or Δ_6–Δ_3 transitions from the character table of the point group C_{6v} which is the group of wave vector Δ, using equation (4.23), we find that these transitions are allowed for $E \perp c$ and forbidden for $E \parallel c$ since the z-component transforms like Δ_1 and x, y, like Δ_5. This explains why the A component in Fig. 6.1 is only present for $E \perp c$. The correspondence, however, holds only for two of the eight equivalent $\langle 111 \rangle$ directions, viz. $\pm\langle 111 \rangle$. The electronic states along the other six directions cannot be derived using the double-zone scheme. It can be assumed that they are transformed to low-symmetry points in the wurtzite zone where the orbital degeneracy of the p-like valence band wave function will be completely lifted. At those points transitions may be allowed for both $E \perp c$ and $E \parallel c$ giving rise to the B component in Fig. 6.1. Phillips[94] proposed transitions at M for $E \perp c$ and K for $E \parallel c$ as the origin of the B component but at present there is not enough knowledge about the states at these symmetry points to give any definite assignment.

The E_1' peak in zincblende is due to L_3–L_3 transitions. The perturbation scheme brings one of the four equivalent L points into Γ and the corresponding Γ_6–Γ_6 transitions are now only allowed for $E \parallel c$. The other three L points again go to points of lower symmetry than Γ where transitions for both polarizations are probably allowed. Thus one should expect a doublet for $E \parallel c$ and a single peak for $E \perp c$. The doublet is unresolved in the measurements but as we mentioned above the position of E_1' lies between 0·12 and 0·28 eV higher for $E \parallel c$ than for $E \perp c$.

The major anisotropy effects can thus be explained in terms of a slightly perturbed cubic lattice. Transitions at Γ are almost unaffected by the transformation from zincblende to wurtzite since there is little interaction between the additional Γ_6 and Γ_3 states and the Γ_5 and Γ_1 states already present at Γ. This also explains the small polarization effects at the absorption edge. For the same reason transitions between the two groups of states have not been observed although they are in principle allowed. Figure 6.3 shows the resulting band structure of CdS along the hexagonal axis according to Cardona and Harbeke.[120] The double group

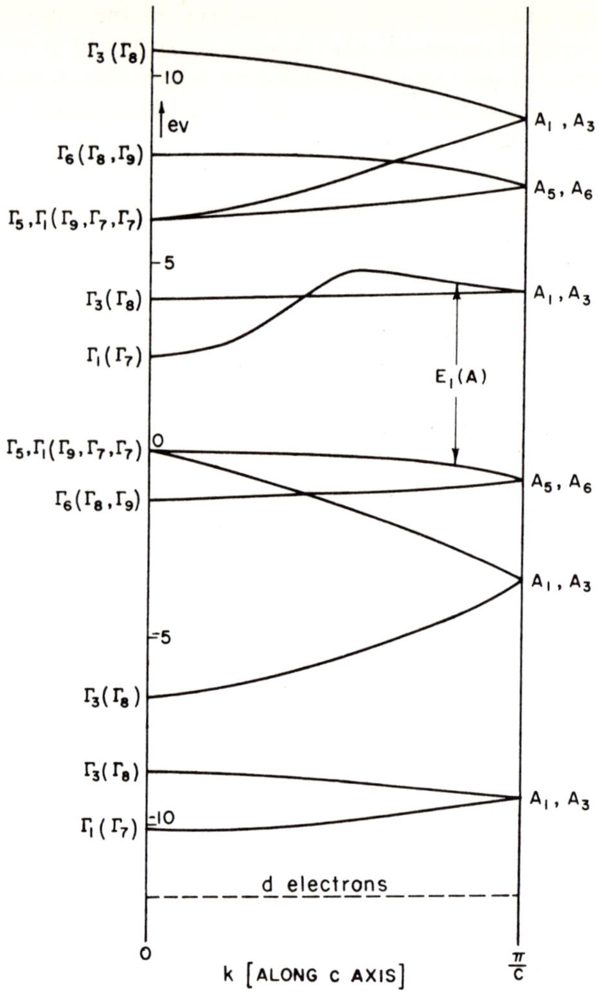

FIG. 6.3. Band structure of CdS along the hexagonal axis (Cardona and Harbeke[120]).

notation at Γ is given in brackets. Γ_1 and Γ_5 states are shown as degenerate because of the small crystal field splitting. Also all *A* states are shown degenerate since the splittings are unknown. The Δ-transitions causing the A-component of E_1 are located between Γ and A in the double zone by analogy with the Λ saddle point in Ge.

Preliminary results of band structure calculations with the OPW method by Collins *et al.*[(124)] seem to indicate that the Γ_5, Γ_1 conduction band state lies about 2 eV higher than is shown in Fig. 6.3. Also pseudopotential[(125)] and APW calculations[(126)] of cubic and hexagonal ZnS place the Γ_{15} (or Γ_5, Γ_1 respectively) conduction band states about 8–10 eV above the valence band maximum. The original assignment of the 5·8 eV peak in ZnS[(120)] has thus to be revised. New experimental data obtained on purely hexagonal ZnS[(126)] also show evidence for a polarization dependent splitting of this peak as in CdS. The 5·8 eV peak is thus probably also caused by Λ_3–Λ_1 transitions in cubic ZnS and by transitions at points along the corresponding Δ symmetry line in hexagonal ZnS.[(125)]

6.2.2. *Bismuth telluride*

We now proceed to crystal structures which are not closely related to the cubic system. As a first example we will treat Bi_2Te_3 and related materials of rhombohedral structure; the point group is D_{3d}. Fig. 6.4 shows the Brillouin zone which is highly compressed parallel to the threefold c-axis. Over the past years

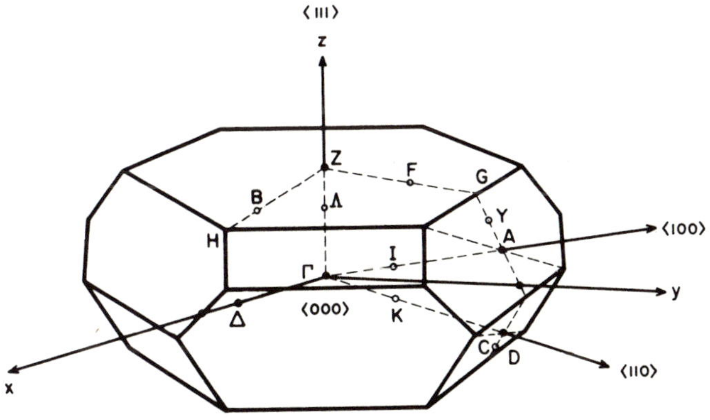

FIG. 6.4. Brillouin zone of Bi_2Te_3.

the multi-valley structure of both the highest valence band and the lowest conduction band in bismuth telluride has been established beyond doubt[(127)]. The valleys are believed to be located at points along ΓA or along ΓD, both directions having equivalent symmetry. Measurements on the optical absorption edge[(128)] indicate indirect transitions and a band gap of 0·15 eV at room temperature. From a symmetry analysis of the Brillouin zone one notes that the groups of the wave vectors Γ, Z (equivalent to Γ), A and D (equivalent to A) contain the inversion. This means that at these points the condition for critical points $|\ _{\mathbf{k}}(E_c - E_v)| = 0$ is fulfilled from symmetry considerations alone. We note further that Γ_3^+, Γ_3^- (and thus also Z_3^+, Z_3^-) and Λ_3 are the only degenerate states in the Brillouin zone, thus only these states can split under spin–orbit interaction. The complete selection rules for all the main symmetry points are given in Table 6.2, both for single and double group representations. The orientation dependence of the selection rules for the states of D_{3d}[60] arises from the fact that the coordinates x and y transform like Γ_3^- whereas z transforms like Γ_1^-. The most interesting aspect is the influence of the spin–orbit interaction on the selection rules. The way in which the states of the double group result from those of the single group was outlined in Section 4.4.2. The state Γ_3, e.g., doubly degenerate in the single group, goes into the doubly degenerate state Γ_6 and the states $\Gamma_4 + \Gamma_5$ degenerate by time reversal. Transitions of the type $\Gamma_3^+ - \Gamma_3^-$ (or $\Gamma_3^- - \Gamma_3^+$ since the parity may be interchanged) allowed for both orientations in the single group, then become possible between two pairs of levels in the form of a quadruplet. The selection rules can now be obtained in the same way as for the single group. One finds two transitions out of four

TABLE 6.2

Selection rules for electric dipole transitions in Bi_2Te_3
($\xrightarrow{\parallel}$ *allowed for* $E \parallel c$, $\xrightarrow{\perp}$ *allowed for* $E \perp c$, $\not\rightarrow$ *forbidden*)

Single group	Double group	Single group	Double group
Γ, Z: $\Gamma_1^+ \xrightarrow{\parallel} \Gamma_1^-$	$\Gamma_4^+, \Gamma_5^+ \xrightarrow{\parallel} \Gamma_4^-, \Gamma_5^-$	A, D: $A_1^+ \xrightarrow{\perp} A_1^-$	$A_3^+ \xrightarrow{\perp} A_3^-$
(D_{3d}) $\Gamma_1^+ \not\rightarrow \Gamma_2^-$	$\Gamma_4^+, \Gamma_5^+ \xrightarrow{\perp} \Gamma_6^-$	(C_{2h}) $A_1^+ \xrightarrow{\parallel,\perp} A_2^-$	$A_3^+ \xrightarrow{\parallel,\perp} A_4^-$
$\Gamma_1^+ \xrightarrow{\perp} \Gamma_3^-$	$\Gamma_6^+ \xrightarrow{\perp} \Gamma_4^-, \Gamma_5^-$	$A_2^+ \xrightarrow{\perp} A_2^-$	$A_4^+ \xrightarrow{\perp} A_4^+$
$\Gamma_2^+ \xrightarrow{\parallel} \Gamma_2^-$	$\Gamma_6^+ \xrightarrow{\parallel,\perp} \Gamma_6^-$		
$\Gamma_2^+ \xrightarrow{\perp} \Gamma_2^-$			
$\Gamma_3^+ \xrightarrow{\parallel,\perp} \Gamma_3^-$			
Λ: $\Lambda_1 \xrightarrow{\parallel} \Lambda_1$	$\Lambda_4, \Lambda_5 \xrightarrow{\parallel} \Lambda_4, \Lambda_5$	I, K: $\Omega_1 \xrightarrow{\parallel,\perp} \Omega_1$	$\Omega_3 \xrightarrow{\parallel,\perp} \Omega_3$
(C_{3v}) $\Lambda_1 \not\rightarrow \Lambda_2$	$\Lambda_4, \Lambda_5 \xrightarrow{\perp} \Lambda_6$	(C_s) $\Omega_1 \xrightarrow{\perp} \Omega_2$	$\Omega_3 \xrightarrow{\perp} \Omega_4$
$\Lambda_1 \xrightarrow{\perp} \Lambda_3$	$\Lambda_6 \xrightarrow{\perp} \Lambda_4, \Lambda_5$	$\Omega_2 \xrightarrow{\parallel,\perp} \Omega_2$	$\Omega_4 \xrightarrow{\parallel,\perp} \Omega_4$
$\Lambda_2 \xrightarrow{\parallel} \Lambda_2$	$\Lambda_6 \xrightarrow{\parallel,\perp} \Lambda_6$		
$\Lambda_2 \xrightarrow{\perp} \Lambda_3$			
$\Lambda_3 \xrightarrow{\parallel,\perp} \Lambda_3$			

possible ones allowed for $E \parallel c$ and three for $E \perp c$. If the splittings of initial and final states are equal, and the ordering of the split components is also the same, then the two $E \parallel c$ transitions coincide and are of the same energy as the central component of the triplet expected for $E \perp c$. Since a considerable spin–orbit interaction can be expected in compounds of such heavy constituents, the appearance of a multiplet structure as discussed above will be a strong indication for transitions between levels of the Γ_3 type. The reflectivity spectrum of Bi_2Te_3[48] for $E \perp c$ in Fig. 6.5 indeed shows two triplets of similar line shape labelled E_1, E_2, E_3 and E_5, E_6, E_7 with identical splittings of about 1·4 eV between the first two components and about 1·0 eV between the

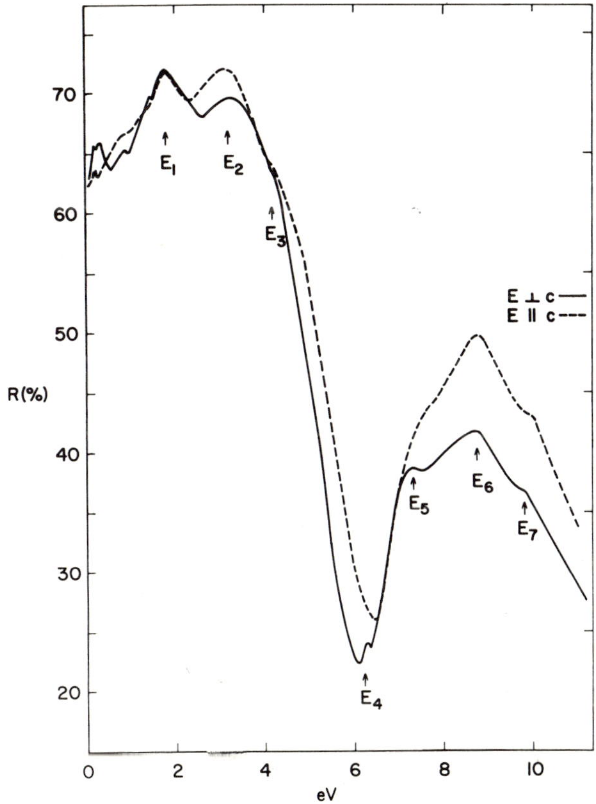

FIG. 6.5. Room temperature reflectivity of Bi_2Te_3 between 0·1 eV and 11 eV for $E \perp c$ and $E \parallel c$ (Greenaway and Harbeke[48]).

last two. The crucial experiment for the verification of the assignment $\Gamma_3^- - \Gamma_3^+$ is the measurement for $E \| c$ and one notes the expected stronger middle components E_2 and E_6. Since the splittings are not equal the $E \| c$ peak should consist of two components 0·4 eV apart which have not been resolved in the experiment. The side bands do not completely disappear for $E \| c$, probably for a number of experimental reasons. Their strength, however, is weakened relatively and it is felt that the $E \| c$ results confirm the assignment given above.

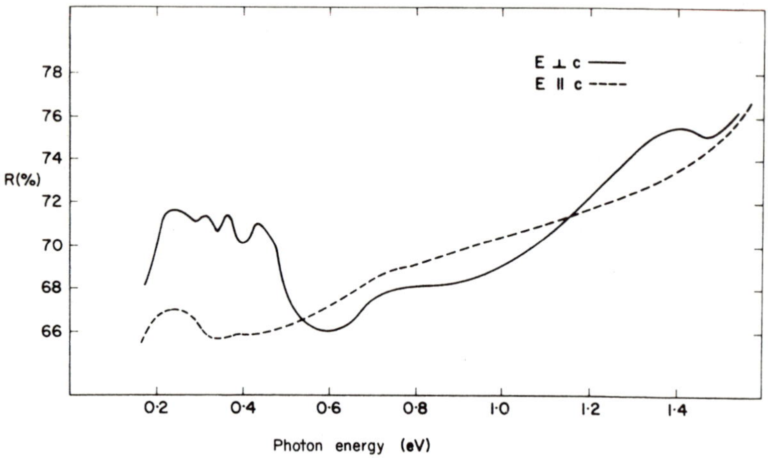

FIG. 6.6. Reflectivity of Bi_2Te_3 at 77°K between 0·18 eV and 1·5 eV for $E \perp c$ and $E \| c$ (Greenaway and Harbeke[48]).

The assignment is also confirmed by the low-energy results which are shown in Fig. 6.6 with an expanded energy scale. Here we find a structured peak around 0·4 eV for $E \perp c$ and another peak at 1·4 eV; both disappear for $E \| c$. We note again the splitting of 1·0 eV which is observed in both triplets. This limits the assignment to transitions from the same spin–orbit split valence band Γ_6^-, $(\Gamma_4^- + \Gamma_5^-)$, to a band Γ_6^+ coming from a nondegenerate band Γ_1^+. The single-group selection rule $\Gamma_3^- \longrightarrow \Gamma_1^+$ is relaxed for the low-energy component $(\Gamma_6^- - \Gamma_6^+)$ of the doublet in the double group, which is also allowed for $E \| c$, but the strength of this transition is not sufficient to make it observable.

E⊥*c* spectra have also been taken of $Bi_2Te_3-Bi_2Se_3$ alloys of different compositions,[(48)] they preserve all the main features of the pure Bi_2Te_3 except that all the transitions shift to higher energy as the selenium concentration is increased. The energies of the peaks E_1 to E_7 as well as the energy gap E_0 are shown in

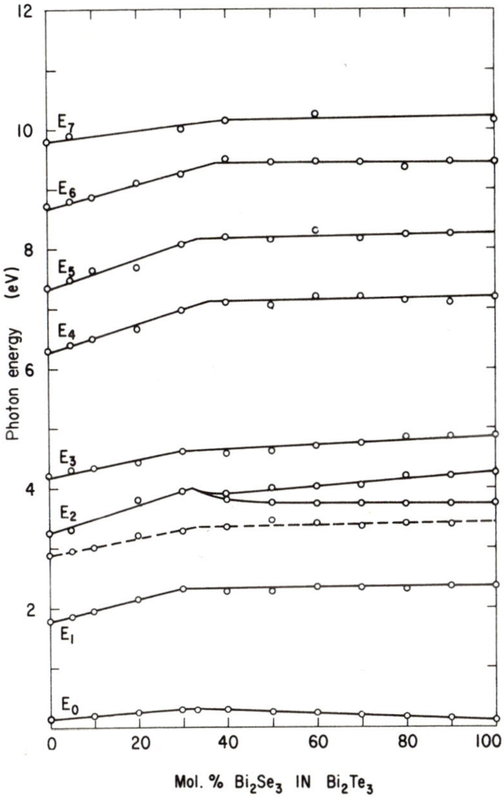

FIG. 6.7. Reflectivity peak energies in Bi_2Te_3–Bi_2Se_3 alloys of different composition (Greenaway and Harbeke[(48)]).

Fig. 6.7. From this behaviour it can be concluded that the band structure of the entire alloy system is qualitatively similar. Furthermore, the nearly identical slopes of the energy versus composition curves for all the peaks E_1 to E_7, and the small difference in slope going from E_6 to E_7 or from E_2 to E_3 indicating the decrease of the spin–orbit splitting of the valence band with increasing addition of selenium, favour the assignments given above. They lead to the energy band model for Bi_2Te_3, shown in Fig. 6.8, in which the main critical points besides the band extrema in the direction A or D are located at Γ or Z in five bands running nearly horizontally, in accordance with the small

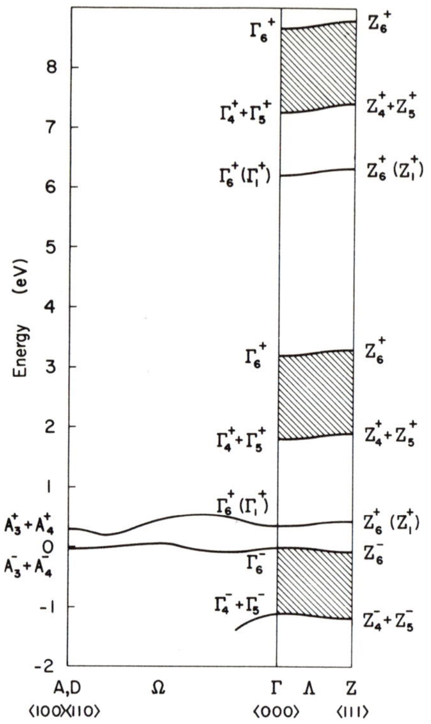

FIG. 6.8. Band structure of Bi_2Te_3 (Greenaway and Harbeke[48]).

group velocity of the electrons in the corresponding c-direction in real space. The **k**-dependence of the higher conduction bands in directions other than ΓZ is not known.

6.2.3. *Selenium*

The point group of hexagonal selenium is D_3. The characteristic feature of the lattice is the occurrence of spiral chains along the c-axis whereby nearest neighbours lie on the same chain. Many attempts have been made to calculate the band structure particularly along the chain direction. The most advanced one was done by Olechna and Knox[129] based on an earlier model of Reitz.[130] They use a tight binding method taking *s–p* hybridization and a screened exchange potential into account, and arrive at a band structure along the k_z direction ΓA as shown in the left half of Fig. 6.9. The conduction bands have been slightly adjusted in order to give a quantitative fit to the reflectivity measurements of Stuke and Keller[131] which

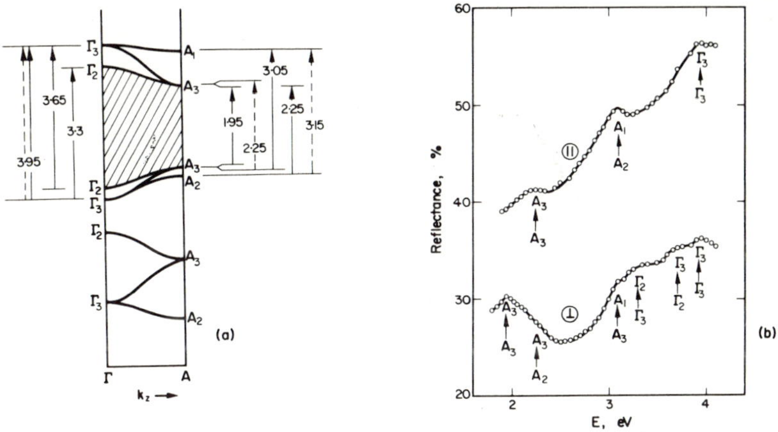

FIG. 6.9. (a) Band structure of selenium along k_z (Olechna and Knox[129]). (b) Room temperature reflectivity of selenium for $E\perp c$ and $E\parallel c$ (Stuke and Keller[131]).

are also shown in Fig. 6.9. Allowed optical transitions, obtained from the group theoretical treatment, are indicated by solid arrows for $E\perp c$ and by dotted arrows for $E\parallel c$. It can be seen from the figure that these assignments give a satisfying explanation of the experimental results obtained with linearly polarized light up to 4·1 eV. The first calculation of the band structure of selenium including other symmetry points and axes (using the Green's Function method of Kohn and Rostoker) was made recently by Sandrock and Treusch.[132] Their results indicate a direct gap at or close to the corner *H* of the hexagonally prismatic Brillouin zone.

6.2.4. *Graphite*

Crystalline graphite of point group D_{6h} is the classical layer substance. Three of the four valence electrons form covalent σ-bonds with the three nearest-neighbour atoms in the same plane. The remaining electron is in a π-state with associated wave functions made up from p_z orbitals. In a two-dimensional band model the π-bands touch at the corners of the two-dimensional hexagon representing the Brillouin zone. Slonczewski and Weiss have developed a three-dimensional model in the tight-binding approximation in the neighbourhood of the zone edges of the Brillouin zone which is then a hexagonal prism. The model has been extensively used to explain the electronic properties of graphite in the region of the Fermi surface.[133] At higher energies ϵ_2 for $E \perp c$ derived from reflectivity measurements[134] shows two distinct maxima near 5 eV and 14 eV. In a recent two-dimensional tight-binding calculation throughout the entire zone, Bassani and Pastori[135] have found that these maxima are due to saddle point transitions at the point Q, the midpoint of the sides of the hexagon, giving rise to logarithmic singularities in the joint density of states of the form (4.13). The first maximum arises from a saddle point in the π-bands and the second one from σ–σ transitions; the selection rules forbid σ–π transitions for $E \perp c$.

6.2.5. *Rutile*

Tetragonal rutile, TiO_2 (space group D_{4h}), is known to be strongly birefringent with values of about 2·4 ($E \perp c$) and 2·7 ($E \parallel c$) for the long-wavelength refractive indices.[136] The fundamental absorption edge at 3·1 eV, on the other hand, shows a very small shift from $E \perp c$ to $E \parallel c$ which cannot account for the birefringence and dispersion. The reflectivity spectra[120] show an exceedingly strong dependence on the direction of polarization. If one deduces the optical constants from the curves, the long-wavelength birefringence and dispersion can be explained. The dispersion curves[136] were fitted with a single harmonic oscillator model with the oscillator energies at 4·38eV for $E \perp c$ and 4·27 eV for $E \parallel c$. The imaginary part of the dielectric constant in Fig. 6.10 deduced from the reflectivity spectra does indeed exhibit one dominant absorption band for $E \parallel c$ at 4·3 eV corresponding to the first strong reflectivity maximum. For $E \perp c$ we find two absorption bands which mainly determine n at long wavelengths. The centre of both bands combined additively is about 0·2 eV higher than the peak position for $E \parallel c$, again in good agreement with DeVore's single oscillator fit of $n_{\perp}(\lambda)$ The dispersion and birefringence of rutile is thus determined by the energy region of highest absorption constant and not by the first absorption edge.

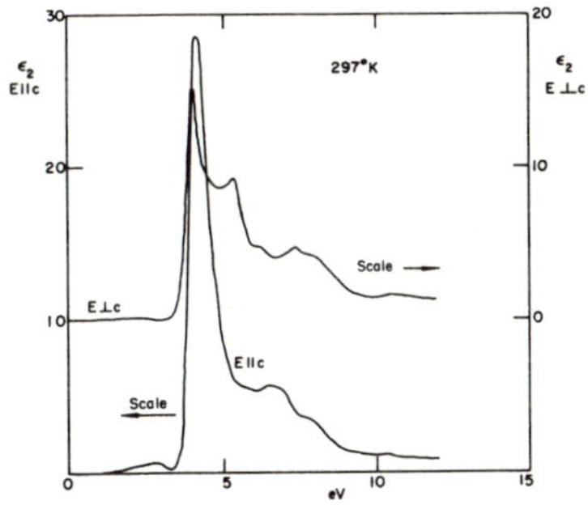

FIG. 6.10. Imaginary part of the dielectric constant of rutile for $E \perp c$ and $E \parallel c$ (Cardona and Harbeke[120]).

This example stresses the interconnection between the low- and high-energy optical constants although only limited conclusions can at present be drawn concerning the band structure. There is a dominant peak in the imaginary part of the dielectric constant around 4–5 eV, particularly for $E \parallel c$. This behaviour has not been encountered in materials of the diamond–zincblende–wurtzite family and indicates a singularity in the joint density of states caused by rather flat conduction and valence bands running nearly parallel, or strong exciton effects. Both of these effects are expected for a strongly ionic crystal like rutile.

6.3. Biaxial crystals

Optically biaxial crystals are considerably more difficult to treat using this approach but the symmetry can again help to clarify the situation. Orthorhombic cadmium antimonide, CdSb, has been investigated by Tauc and Abrahám[137] and Fig. 6.11 shows R, ϵ_1 and ϵ_2 up to 1·5 eV for the three different orientations of the electric vector parallel to the a-, b- and c-axis. The optical constants behave entirely differently for all three directions. The use of the selection rules in the analysis of the data is based upon the fact that certain combinations of allowed and forbidden transitions for the three polarizations can only take place at certain points in **k**-space thus excluding all other points for an assignment of the structure. As in the case of bismuth telluride there are transitions for which the more restrictive single group selection rules are relaxed by spin–orbit interaction.

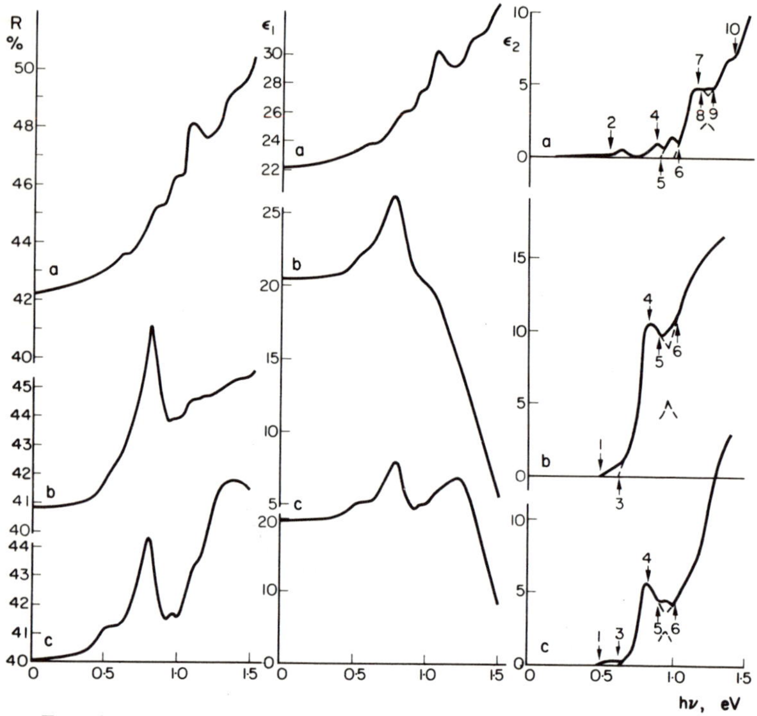

FIG. 6.11. Reflectivity R and real and imaginary parts of the dielectric constant ϵ_1, ϵ_2 of CdSb for (a) $E \parallel a$, (b) $E \parallel b$, and (c) $E \parallel c$ (Tauc and Abrahám[137]).

Again, as far as the interaction can be considered as a weak perturbation the selection rules without spin will hold approximately in the spin case; because of the comparatively small matrix elements one is not able to distinguish between a completely and an approximately forbidden transition in the experimental curve. This procedure has proved to be effective for the discussion of the interband transitions in CdSb. The number of possible locations for the observed interband transitions has been restricted to a few points in each case. Together with results on the transport properties of CdSb and the reasonable assumption that the absolute band extrema are not caused by spin–orbit splitting, one can reduce the number of possible locations for these band extrema further, and arrive at a possible structure for the highest valence band and the lowest conduction band.

CHAPTER 7

DEFORMATION PHENOMENA

7.1. Hydrostatic pressure

Measurements of the optical properties, mostly absorption, of diamond and zincblende semiconductors under hydrostatic pressure have proved useful in identifying the symmetries of the lowest conduction bands. The results show that the pressure coefficients of the energies of the Γ_2' (or Γ_1 in zincblende), L_1 and Δ_1 states, with respect to the Γ_{25}' (or Γ_{15}) top valence band states, are about 12×10^{-6} eV kg^{-1}cm^2, 5×10^{-6} eV kg^{-1}cm^2 and -2×10^{-6} eV kg^{-1}cm^2 respectively. These values are independent of the material.[88,138] This characteristic behaviour is a consequence of the different symmetries of the wave functions associated with the three states. Since the valence electron wave functions are orthogonal to the core wave functions the valence electron eigenstates depend strongly on the core states, particularly on the energy difference between the highest core state and the lowest valence state. Herman[139] has shown that the change of this parameter, the so-called "core shift", results in a different response of the energy levels depending on the symmetry of the corresponding wave functions. States with *s*-like symmetry are far more core shift dependent than *p*-like states since the probability of *s*-functions is very high close to the core. The changes of the energy levels with hydrostatic pressure occur because the interatomic distances decrease, whereas the core can be regarded as essentially incompressible. The consequence is a rearrangement of the valence wave function relative to the core functions, in other words a core shift.

The variation of the important energy levels in Ge with lattice constant a has been calculated by Bassani and Brust[140] from the variation of the pseudopotential parameters with lattice constant. The results (relative to Γ_{25}') are shown in the right half of Fig. 7.1 and can be seen to agree qualitatively with the experimental data.

The pure s-like state Γ_2' varies most strongly with a; the state L_1, being somewhat less s-like, shows an intermediate slope, whereas all other levels without s-admixture vary approximately in the same manner as Γ_{25}'. A similar behaviour has been found for the variation of the energy levels with lattice constant in Ge–Si alloys, calculated from a linear interpolation of the pseudo-

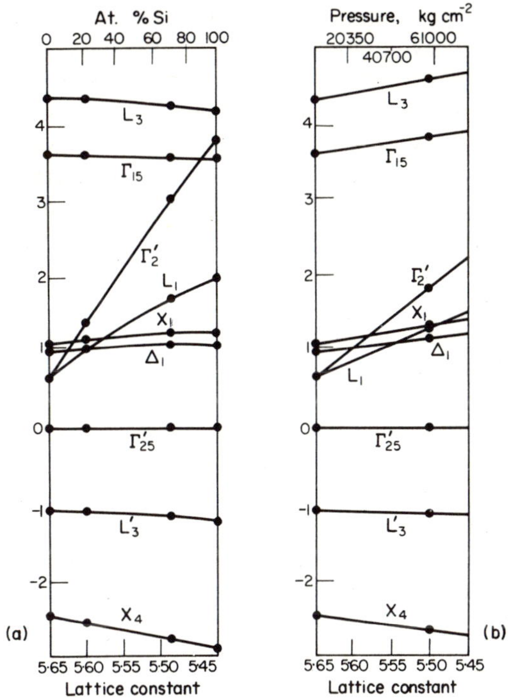

FIG. 7.1. Behaviour of the energy levels as a function of (a) the silicon concentration in Ge–Si alloys, and (b) hydrostatic pressure in Ge (Bassani and Brust[140]).

potential parameters of Ge and Si (see Fig. 7.1). The variation for the same change in lattice constant is much stronger than with pressure because of the admixture of other core functions. The consequences of the strong sensitivity of the s-levels are the differences in the band structure of Ge and Si. The lowest $\mathbf{k} = 0$ conduction band state in Ge, Γ_2', lies about 3 eV higher in Si and thus above Γ_{15}; also the lowest conduction band L_1 has moved to much higher energies so that Δ_1 becomes lowest in Si. On the basis of this model it has also been possible to explain the absorption edge data[141] and the reflectivity measurements in Ge–Si alloys.[72] Another consequence of the core-shift sensitivity, this time in conjunction with the spin–orbit interaction, is that Γ_2' lies below Γ_{25}' (3/2) in α-Sn.[80] From the $\mathbf{k}\cdot\mathbf{p}$ interaction with one of the two bands degenerate at Γ_{25}' (3/2), Γ_2' obtains a downward curvature off $\mathbf{k}=0$ whereas the band to which it is coupled via the interaction is curved upwards. This brings the Fermi level in coincidence with Γ_{25}' (3/2) so that the energy gap is zero at all temperatures making α-Sn the ideal semimetal.

The pressure coefficients of the energies of reflectivity maxima have also been measured in a few cases. Zallen, Paul and Tauc.[142–3] found a coefficient of $(7{\cdot}2\times10^{-6}\pm10\%)$ eV kg^{-1}cm^2 for the first reflectivity maximum in Ge, InAs and GaSb, caused by Λ_3–Λ_1 transitions. Since L_3' and thus also Λ_3 vary like Γ_{25}' with pressure (see Fig. 7·1) and since the band Λ_1 connects Γ_2' with L_1, one would indeed expect a value for the pressure coefficient of $\Lambda_3-\Lambda_1$ between those of $\Gamma_{25}'-\Gamma_2'$ and $\Gamma_{25}'-L_1$. The pressure coefficient of the first reflectivity maximum in Si is 5×10^{-6} eV kg^{-1}cm^2[143] indicating again the different nature of the transitions. The value for the pressure shift of the X_4–X_1 peak in Si of about 3×10^{-6} eV kg^{-1}cm^2 is in reasonable accordance with the model of Bassani and Brust.

The different pressure coefficients are also reflected in the linear temperature coefficients of critical point energies. The latter can be written[144] in simple cases as:

$$\left(\frac{\partial E}{\partial T}\right)_P=\left(\frac{\partial E}{\partial T}\right)_V-\frac{\alpha}{\beta}\left(\frac{\partial E}{\partial P}\right)_T, \tag{7.1}$$

where α is the volume thermal expansion coefficient and β the volume compressibility. The first term is due to electron–phonon interaction, the second term is due to the dilation of the lattice and is thus proportional to the pressure coefficient $(\partial E/\partial P)_T$. In diamond and zincblende materials both terms have the same sign and are about equal. This makes it plausible that the temperature coefficients of reflectivity peaks due to Λ_3–Λ_1 transitions are $-(4{\cdot}7\pm0{\cdot}8)\times10^{-4}$ eV deg^{-1} whereas the 3·4 eV maximum in Si shifts by $-(2{\cdot}7\pm0{\cdot}6)\times10^{-4}$ eV deg.$^{-1}$[145]

Zallen and Paul[91] have been able to identify a shoulder in the reflectivity spectrum of GaP at 2·8 eV by its pressure coefficient of $(10{\cdot}7 \pm 15\%) \times 10^{-6}$ eV kg^{-1}cm^2 as a Γ_{15}–Γ_1 transition. The situation at the first strong reflectivity maximum in GaP at 3·7 eV is less clear. The pressure and temperature coefficients of $5{\cdot}8 \times 10^{-6}$ eV kg^{-1}cm^2 and $-2{\cdot}5 \times 10^{-4}$ eV deg^{-1} seemed to indicate that the peak is caused by Γ_{15}–Γ_{15} transitions or close to Γ in the $\langle 100 \rangle$ direction.[146] Recent reflectivity measurements on GaAs–GaP alloys[147–8] as well as electroreflectance data[92,149] show that these transitions occur at 4·8 eV instead. If this assignment is made, the origin of the 3·7 eV peak is not clear, also the alloy studies do not give conclusive evidence concerning its nature.

7.2. Piezoreflectance

7.2.1. *Germanium*

The optical properties of strained crystals are again described by symmetric second-rank tensors n and k or ϵ_1 and ϵ_2 like those of the anisotropic crystals discussed in Section 6.1. Particularly useful are measurements in cases where the symmetry of the crystal is lowered by the application of stress so that the stress-induced differences can be discussed in terms of the lower symmetry and the appropriate selection rules. For a cubic crystal under an uniaxial deformation, $\Delta\epsilon_2$ can be written as

$$\Delta\epsilon_2 = \begin{pmatrix} \Delta\epsilon_{2\perp} & 0 & 0 \\ 0 & \Delta\epsilon_{2\perp} & 0 \\ 0 & 0 & \Delta\epsilon_{2\parallel} \end{pmatrix} \tag{7.2}$$

if the third axis is taken parallel to the strain axis. The stress induced changes $\Delta\epsilon_{2\perp}$ and $\Delta\epsilon_{2\parallel}$ may be derived by Kramers–Kronig analysis from the variations of the reflectivity $\Delta_{R\perp}$ and $\Delta_{R\parallel}$ measured with linearly polarized light perpendicular and parallel to the stress axis.[150] For small strains the change in reflectivity is mostly taken directly proportional to $\Delta\epsilon_2$. Piezoreflectance measurements on Ge under static deformation have been reported by Philipp, Dash and Ehrenreich[37] and Gerhardt.[36,151] The technique used (see Section 3.3.5.) allows measurements to be taken under equivalent extension and compression on opposite sides of the sample at the same time. Figure 7.2 shows the reflectivity of uniaxially stressed crystals at the first doublet peak in Ge (Λ_3–Λ_1 transitions) for a strain component along the stress axis of 0·4%.[151] For stress along the ⟨001⟩ axis there is a polarization independent shift of the peaks. This is to be expected since all ⟨111⟩ bands are shifted equally by ⟨001⟩ stress and thus no polarization dependent splitting of the energies occurs. Stress along the ⟨111⟩ axis, however, divides the eight equivalent ⟨111⟩ axes into two groups of two and six axes.

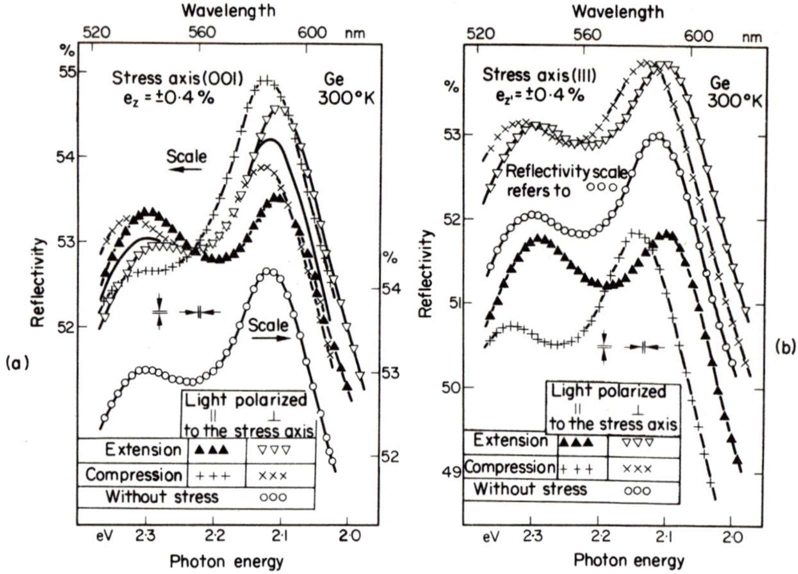

FIG. 7.2. 300°K reflectivity of unstressed and uniaxially stressed Ge between 2·0 and 2·4 eV (Gerhardt[151]).

We have thus exactly the same situation as for Λ-transitions in wurtzite materials discussed in Section 6.2.1. and the selection rules of point group C_{3v} given in Table 6.2 are valid. Probably due to the small deformation, one does not observe the splitting of the peaks as in wurtzite CdS and CdSe, but the experiment shows a marked polarization dependence of the stress-induced energy shift (see Fig. 7.2(b)). The results for both stress axes thus confirm the assumed assignment. From the measured energy shifts we can determine the deformation potential constants $E_{1c}^{\Lambda} - E_{1v}^{\Lambda} = -5{\cdot}7\ \text{eV} \pm 5\%$ in full agreement with the value obtained from measurements under hydrostatic pressure,[142-3] and $E_{2c}^{\Lambda} = 5\cdot1$ eV $\pm 20\%$. The experiment also shows a strain induced change in the magnitude of the peaks. For $\langle 001 \rangle$ stress the 2·1 eV peak increases and the 2·3 eV peak decreases for E parallel to the stress axis, and vice versa for the perpendicular orientation. It can be shown[151] that the reduction in crystal symmetry at the line Λ from C_{3v} to C_s also accounts for the changes in magnitude of the two peaks.

Polarization effects on the 4·5 eV peak[36] are less pronounced due to its mixed origin of X_4–X_1 and Σ_4–Σ_1 transitions. Preliminary data for the piezoreflectance of unpolarized light between 1·6 and 4·6 eV have also been obtained with the dynamic technique.[38]

7.2.2. *Silicon*

Static piezoreflectance measurements of Si around the 3·4 eV maximum[151] show that for $\langle 001 \rangle$ stress the sign of the energy shift with compression depends on the polarization, whereas $\langle 111 \rangle$ compression always results in a positive shift. This indicates that some contribution to the peak can result from transitions occurring at points along the $\langle 100 \rangle$ axis, most probably Δ_6–Δ_6 and Δ_7–Δ_6 (Δ_5–Δ_1 in the single group notation). The

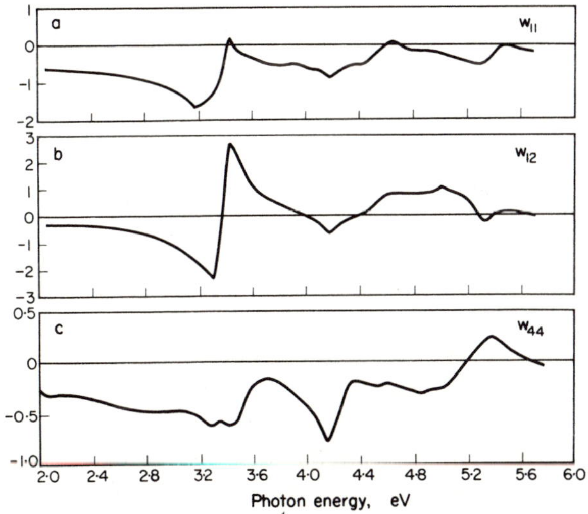

FIG. 7.3. Fundamental components W_{11}, W_{12} and W_{44} of the piezoreflectance tensor of Si (Gobeli and Kane[39]).

observed broadening for the perpendicular orientation is explained by a strain-induced splitting of the valence bands Δ_6 and Δ_7. The amount of line broadening is equal to the strain-induced splitting of the $\Lambda_8(\Lambda'_{25})$ band found by cyclotron resonance.[152] This implies that the Δ critical point is close to Γ as predicted by detailed pseudopotential calculations.[146] It should be noted that by application of stress not only the energy but also the position of critical points may change along symmetry lines which in turn may cause an additional change in energy.

Gobeli and Kane[39] measured the dynamic piezoreflectance between 2·0 and 5·8 eV and discussed the results in terms of the fourth-order piezoreflectance tensor W which connects the second-rank reflectivity tensor to the second-rank strain tensor u_{kl}. In the case of cubic symmetry W has three independent components W_{11}, W_{12} and W_{44}. Figure 7.3 shows these three fundamental piezoreflectance constants measured by Gobeli and Kane. For a critical point which does not split in the strain field the components of W take just the form of the derivative of ϵ_2. Thus for an M_0 critical point (see Table 4.1) and correspondingly for M_2 the nonanalytic contribution to ϵ_2 yields

$$\frac{d\epsilon_2}{du_{kl}} = -\frac{c_0}{2}\frac{d\omega_0}{du_{kl}}(\omega-\omega_0)^{-1/2}; \quad \omega > \omega_0 \tag{7.3}$$

and from the differentiation of the Kramers–Kronig relation

$$\frac{d\epsilon_1}{du_{kl}} = -\frac{c_0}{2}\frac{d\omega_0}{du_{kl}}(\omega_0-\omega)^{-1/2}; \quad \omega < \omega_0. \tag{7.4}$$

The two quantities are equal but occur on different sides of the critical point energy. They contribute with different weighting factors to ΔR. The corresponding equations for M_1 and M_3 show that in this case both terms have opposite sign. The shape of W_{12} around 3·4 eV thus gives evidence that the singularity is of the type M_1 or M_3; M_3 is excluded by the ϵ_2 data. W_{11} is of similar shape but it should also be noted that the dynamic technique can give an energy independent error signal so that the true zero may not be identical with the zero line in Fig. 7.3(a). The much weaker peaks in W_{44} in this region again provide evidence that the 3·4 eV peak arises to some extent from a Δ critical point since its contribution to W_{44} will be zero.

CHAPTER 8

EXCITONS

8.1. General

In the previous sections on the band structure of isotropic and anisotropic materials we described mainly the scattering continuum of direct interband transitions in terms of the one-electron approximation; little mention was made of exciton effects. We now turn to these, and give particular attention to recent results on metastable excitons which are degenerate with a scattering background and which have been found to be present in most semiconductors and insulators. The original concept of excitons was introduced by Frenkel[(153)] who considered the exciton as an excited state of an atom or molecule. The rather weak interaction with other atoms allows the excitation to move through the crystal in a type of diffusive hopping process. The Frenkel model is useful for describing exciton effects in molecular crystals where the exciton is effectively localized on a small number of molecules, but cannot be used to describe the behaviour in semiconductors. In these substances a better description results from the Wannier model[(154)] in which the exciton is composed of wave packets made out of hole valence band states and electron conduction band states. For the normal parabolic band edge excitons in materials with high dielectric constant (e.g. semiconductors), only states lying very near to the band extrema are involved, and under these conditions the exciton causes a hydrogen-like series of bound states lying directly below the direct interband threshold. Elliott[(155)] has used the Wannier model to derive the quantitative behaviour of excitons for direct and indirect, and allowed and forbidden M_0 transitions. The result for allowed direct transitions shows that the exciton produces a series of absorption lines at energies given by:

$$\hbar\omega = E_0 - \frac{G}{n^2}, \tag{8.1}$$

where E_0 is the energy of the interband threshold, G the binding energy of the exciton and n an integer. The intensity of the lines decreases in proportion to n^{-3}. As the absorption edge is approached the lines overlap to give a continuum which is independent of energy; the interband absorption which rises in proportion to $(\hbar\omega - E_0)^{1/2}$ when $(\hbar\omega - E_0) \gg G$ (see Section 4.3) is considerably modified in the threshold region. For direct forbidden transitions an identical line series is predicted except that the first line is missing; the intensity should decrease in proportion to $(n^2 - 1)/n^5$ and the oscillator strength of the lines will be lower. Elliott's model provides a good quantitative description of the excitons associated with the direct allowed absorption edges in Ge,[156–7] GaAs,[157] and GaSe[158] and with the direct but forbidden absorption edge in Cu_2O.[159–60]

Recently it has been realized that there exists a second class of exciton resonance which plays an important role in the band structure and optical properties of semiconductors and insulators. The resonances have been variously given the name hyperbolic, saddle point, metastable or extra excitons. Just as the normal parabolic excitons are derived from M_0 interband thresholds, so the extra resonances are associated with a scattering structure from higher energy bands (for example, from saddle point transitions of type M_1 or M_2). In principle, exciton states should be associated with every critical point since the condition for the existence of a critical point $|\nabla_k (E_c - E_v)| = 0$ is identical with the condition for an exciton that electron and hole have the same group velocity. Any model used to describe the extra excitons must account for the experimentally observed line shape and also explain how such exciton effects can exist degenerate with a strong scattering continuum of interband transitions. There is no rigorous theoretical calculation of bound exciton states at saddle points yet. Their existence, originally conjectured by Phillips,[69] has been questioned by Duke and Segall.[161] Velický and Sak[162] have treated the electron-hole interaction at critical points in two limiting cases. Firstly for a short-range Koster–Slater-type of potential, the line shape of ϵ_2 at M_0 and M_1 critical points is enhanced and sharpened whereas it is smeared out and weakened at M_2 and M_3 type points. Secondly, for a long-range Coulomb interaction, resonant lines are only found for M_0 points; however, for the case of a very prolate M_1 saddle point, with a large negative value for m_3, one obtains a singularity of ϵ_2 as an asymptotic property as $m_3 \rightarrow \infty$. This latter situation may hold for many M_1 critical points or extended regions (see Section 5.2) so that a modification of the one-electron M_1 line shape due to electron–hole interaction is by no means unrealistic. The explanation of extra excitons that follows is due mainly to Phillips who has carried out a detailed analysis of these effects in diamond and zincblende semiconductors, the alkali halides, and the solid rare gases.[69,110–11,113,163–4] Because the extra excitons are degenerate with a scattering background they cannot form stable bound states. The long lifetime (and hence strength) is due to their decay mechanism which is one of autoionization. Autoionization arises from elastic, energy conserving transitions which can take place when states of different configurations—in this case states belonging to a discrete spectrum and states belonging to a continuous spectrum—are mixed together. The transitions involve two electrons simultaneously and are thus slower than direct one-electron transitions. On this basis we can qualitatively explain the large strength of some metastable excitons

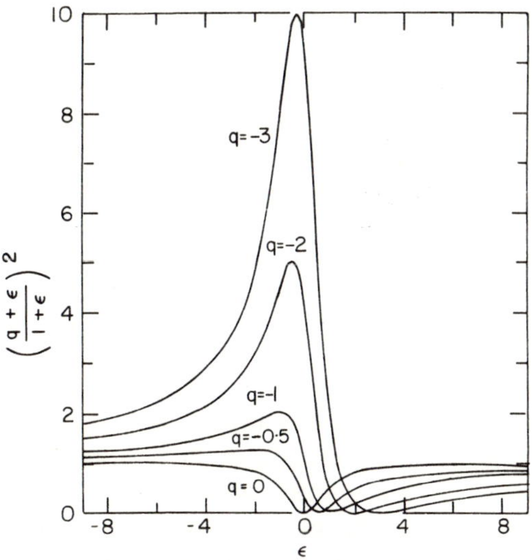

FIG. 8.1. Line shape of an exciton resonance showing the asymmetry and antiresonance caused by the interference between the resonance and the scattering background (Fano[165]).

which is observed experimentally. The effects of configuration interaction are well known in atomic spectra, and Fano[165] has given a detailed discussion of the autoionization of the $2s2p\,^1P$ resonance of atomic helium which is observed at 60 eV in inelastic electron scattering experiments. The situation here is closely analogous to the model that is proposed for the extra excitons in semiconductors and insulators. The basic results of Fano's analysis for the situation in He are shown in Fig. 8.1. The parameter ϵ plotted horizontally is a reduced energy variable defined by

$$\epsilon = \frac{2(\hbar\omega - E_0 - E_1)}{\Gamma}, \tag{8.2}$$

where E_0 is the unperturbed energy of the exciton, and E_1 is the shift of the real exciton resonance with respect to E_0 caused by the interaction with the continuum. Γ indicates the spectral line width of the autoionized state. Clearly ϵ corresponds to the energy scale in an optical or electron loss experiment, with $\epsilon = 0$ indicating the position of the resonance. The vertical scale in Fig. 8.1 is the function $(q+\epsilon)^2/(1+\epsilon^2)$ where q^2 is proportional to the ratio of the transition probability to the modified exciton state and the transition probability to a band width Γ of the unperturbed continuum states. This function would be represented by the imaginary part of the dielectric constant in an optical experiment. Curves are shown for several negative q values. Changing the sign of q simply reverses the horizontal scale and gives a resonance shape with the steep portion towards lower energies. Both situations are physically possible. Large values of q of either sign indicate that a large fraction of the total oscillator strength in that region has been transferred to the exciton resonance. The 60 eV resonance in He can be accurately fitted to the behaviour shown in Fig. 8.1: q is obtained from the experimental data and is found to be negative; Γ is treated as an adjustable parameter and in fact turns out to be about 0·5% of the excitation energy. Several variations of the ideal behaviour of Fig. 8.1 are possible in real materials. As stated above, q can be either positive or negative, the negative sign is characteristic of resonances generated by attractive interactions and is the situation we would expect in semiconductors and insulators when a single exciton resonance is mixed with an approximately constant scattering background. We can also consider two or more closely spaced exciton resonances which interfere with each other and mix with a scattering continuum. In this case a series of alternating resonances and antiresonances will result but two other effects can also take place. In general there will be a modification of binding energies of the excitons, also the sign of q may be reversed. When measuring, for example, absorption coefficients, the effect of two close excitons on an approximately constant scattering background is to give two peaks separated by a symmetrical antiresonance which goes below the level of the absorption due to the scattering continuum. In ideal materials the antiresonant dips both for single and multiple excitons mixed with a continuum should drop to zero. In real crystal spectra this situation is rarely observed owing to a number of inhomogeneous broadening effects which tend to decrease the antiresonances. Interaction with the lattice either through phonons or defects will reduce the strength of the antiresonant dip compared with the resonant peak. The interaction of excitons of different configurations with the scattering continuum is also favourable for Raman emission of photons (induced by completely inelastic scattering between the resonances) corresponding to the energy difference between the resonances which also reduces the antiresonances. Toyazawa *et al.*[166] have shown that the coexistence of discrete and continuum states can also lead to a reversal of a singularity in the ϵ_2 spectrum, i.e. the appearance of a downward cusp at the energy of an M_1 type critical point.

From the above considerations we see that in order to investigate extra excitons in the fundamental absorption region of real materials the measurement should be made at as low a temperature as possible to avoid serious broadening effects. As with all exciton measurements crystal perfection is also important. We should study the line shape of structure in either the reflectivity or absorption spectra and also if possible in the imaginary part of the dielectric constant; the features we look for are asymmetries of the type illustrated in Fig. 8.1 or symmetrical dips between maxima which go below the average level of the interband continuum. In materials with low static dielectric constant and high exciton binding energy (e.g. the alkali halides) where complicated exciton structure is present we may expect to observe differences in the binding energy of excitons associated with spin–orbit split thresholds due to interference effects. In more nonpolar materials such as the diamond and zincblende semiconductors exciton binding energies are very small, and any exciton effects will certainly be degenerate with structure due to interband transitions. In these cases the behaviour of the line shape as a function of temperature becomes an important factor in recognizing exciton effects as exciton-dominated transitions will sharpen much more rapidly than interband transitions as the temperature is decreased.

Before discussing experimental exciton spectra we deal briefly with the nature of exciton effects in polar and nonpolar materials. Strong exciton effects are not common in the more nonpolar semiconductors at energies well above the fundamental absorption edge although much more common than was realized a few years ago; structure in the optical constants is caused mainly by interband transitions at critical points in momentum space, and exciton effects as we shall see in the following section result in sharpening interband edges and thresholds. These materials are characterized by small lattice parameters and very large static dielectric constants. In the alkali halides exciton effects dominate the low-energy fundamental region and the critical points associated with them cause only weak interband structure. The explanation of this large difference is associated with the large lattice parameters and small dielectric constants of the alkali halides. A small dielectric constant implies a correspondingly small shielding of the Coulomb interaction. If the Coulomb interaction is introduced into the general Hamiltonian of the system in the correct manner then excitons appear as intrinsic excitations and are possible eigenstates of the system. One obtains direct Coulomb terms together with exchange terms. The latter predominate and cause an attractive Coulomb interaction between electron and hole, i.e. the exciton. The exchange terms should be screened by the dielectric constant, consequently one expects much larger exciton effects in ionic materials than in nonpolar materials, and this is indeed found experimentally. This qualitative picture also fits in with the diamond and zincblende materials as there is clear experimental evidence that metastable exciton effects are stronger in the II VI materials than in germanium. We shall also see in Section 8.4 that intermediate cases between the covalent semiconductors and the alkali halides exist. For example, in CdI_2 we find strong extra exciton effects degenerate with a strong interband scattering continuum.

8.2. Excitons in diamond and zincblende materials

Parabolic M_0 excitons associated with the lowest direct threshold have been seen in many of the diamond and zincblende materials. As we saw in Section 8.1 the band edge excitons in Ge and GaAs can be adequately explained using Elliott's model. Macfarlane and Roberts[(167)] have identified an exciton associated with the lowest 1·1 eV indirect Γ'_{25}–Δ_1 transition in silicon. Band edge excitons associated with the M_0 direct thresholds in ZnTe and hexagonal CdS have been measured by Hopfield and Thomas[(168)]; the interpretation of these excitons requires the effect of spatial dispersion to be taken into account.

Metastable excitons are present in the region of both the main near-ultraviolet interband transitions in the more polar zincblende semiconductors; in fact the first indication that excitons might be present around the E_1 transition came from absorption data on evaporated films of CdTe[(169)] where the expected spin–orbit split thresholds were shown to sharpen into peaks at low temperature. Metastable exciton effects associated with both the E_1 and E_2 saddle-point maxima can be clearly seen by referring to the reflectivity curves of ZnTe shown in Fig. 5.4. The line shape of the E_1 doublet is determined primarily by the presence of a saddle-point exciton, and only to a lesser extent by the $\Lambda_3 - \Lambda_1$ scattering transitions with which it is degenerate. The peaks show the characteristic asymmetry and expected strong temperature dependence of extra excitons. The E_2 peak due to nearly degenerate X_5–X_1 and Σ_4–Σ_1 scattering transitions (of type M_1 and M_2 respectively) sharpens due to exciton formation; again the peak shows slight asymmetry at low temperature and a strong temperature dependence of the line width. Metastable excitons associated with the E_1 and E_2 transitions are clearly resolved in most of the II VI semiconductors. In particular Phillips[(69)] has estimated values of $q = -4{\cdot}5$ and $\Gamma = 0{\cdot}016$ eV for the E_1 resonance in CdTe. The high value of q indicates the predominance of the metastable exciton over the Λ_3–Λ_1 scattering background. Also in line with the result for the $2s2p\,^1P$ helium resonance, the decay width Γ is about 0·5% of the excitation energy. As one goes from the II VI materials to the less polar III V and group IV semiconductors the saddle-point resonances become weaker and harder to resolve, which agrees with the qualitative argument given in Section 8.1. Clearly the q values are much less, and the main effect of the weaker resonances is to cause asymmetric line narrowing as the temperature is decreased. Such effects can still be seen in the reflectivity of InAs and InSb[(170)] for both the E_1 and E_2 peaks. In Ge extra excitons have not been detected so far in reflectivity, but they have been found in ϵ_2 spectra determined from the transmission of single crystals. In Fig. 8.2 we show ϵ_2 curves at 77°K and 300°K for Ge

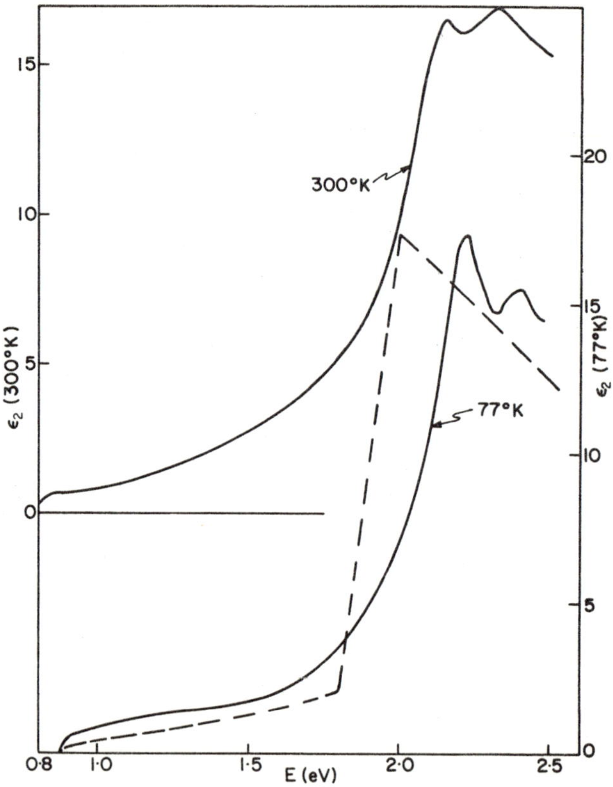

FIG. 8.2. Experimental (solid) and theoretical (dashed) ϵ_2 curves for Ge in the neighbourhood of the E_1 reflectivity maximum (Harbeke[(171)]).

in the region of the E_1 doublet.[171] The 77°K curve shows the expected exciton line sharpening and the deviation from the theoretical one-electron line shape of the M_1 type critical point.

In diamond and silicon somewhat different situations exist. As we pointed out in Section 5.3, the E_1 peak in Si is probably caused by several nearly degenerate transitions. The degeneracy of an M_0 transition with an M_1 type transition would result in a line shape which is approximately a step function, a situation favourable for exciton formation. It has been suggested[172] that a strong exciton (referred to as a hybrid exciton) is in fact present at the E_1 peak in Si. The experimental reflectivity data (Fig. 5.2) support this contention. The lowest direct gap in diamond is due to Γ'_{25}–Γ_{15} transitions. Reflectivity data by Clark *et al.*[173] showed a strongly temperature-dependent peak at 7 eV which Phillips[174] attributed to hybrid excitons at Γ partly associated with the Γ'_{25}–Γ_{15} gap at 8·7 eV. More recent measurements by Roberts *et al.*[175] show a 7·2–7·6 eV doublet without antiresonance effects and no structure at 8·7 eV which casts some doubt on the hybrid exciton interpretation. These authors place the M_1, Γ'_{25}–Γ_{15} edge at 7·2 eV in good agreement with pseudopotential band structure calculations by Saslow *et al.*[83]

8.3. Excitons in the alkali halides

Because of the much weaker dielectric screening, the exciton binding energies in the alkali halides are large compared with the values of 0·01–0·1 eV expected in diamond and zincblende materials. Thus exciton resonances will in general be well separated from the parent interband edges and should be easier to analyse. As we saw in Chapter 5 the absorption (and reflectivity) spectra of most alkali halides are dominated by the exciton transitions associated with the lowest spin–orbit split M_0 thresholds at the centre of the Brillouin zone. The experimental curves for KBr shown in Fig. 5.15 typify this behaviour. In fact these parabolic excitons can provide the simplest examples of interference of an exciton with continuum states. Although the $\Gamma(3/2)$ exciton may lie below the fundamental absorption edge, background absorption can be generated by defects or phonons. This is illustrated in Fig. 8.3 which shows reflectivity data for KBr at 77°K and 300°K in the 6–9 eV region. The strongly asymmetric 6·77 eV $\Gamma(3/2)$ exciton line at 77°K is followed by an antiresonance which goes below the average level of the scattering background. The reflectivity at the maximum is less than 2% whereas a value of about 5% is expected from the static dielectric constant alone, without any other contributions. The energies of the $\Gamma(3/2)$ and $\Gamma(1/2)$ excitons for the face-centred cubic structure alkali halides are given in Table 8.1.

According to Phillips the occurrence of metastable excitons associated with L_3'–L_1 transitions is demonstrated in the 80°K absorption spectrum of KI shown in Fig. 8.4. The Γ_{15}–Γ_1 and the

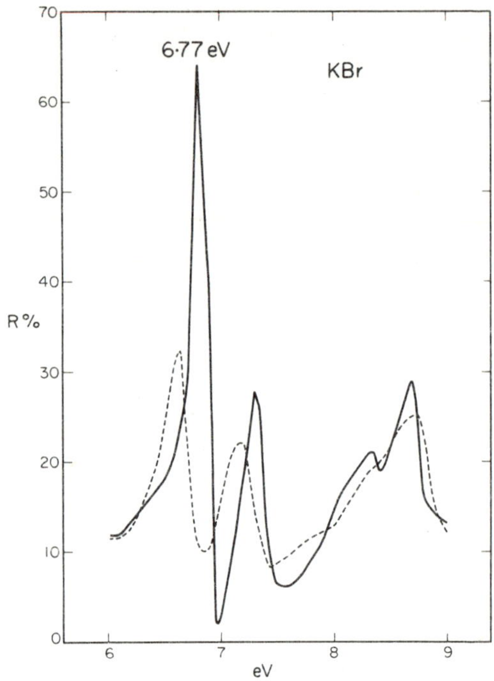

FIG. 8.3. Reflectivity of KBr near the absorption edge. Solid curve 77°K, dashed curve 300°K (Greenaway and Nitsche[177]).

metastable L_3'–L_1 excitons are both resolved with the characteristic spin–orbit splitting of the parent Γ_{15} and L_3' states. The L excitons are weaker than the Γ excitons, and this is true for all the alkali halides. The reverse is observed in the spectrum of solid xenon[111] where the strong L point excitons are separated by a

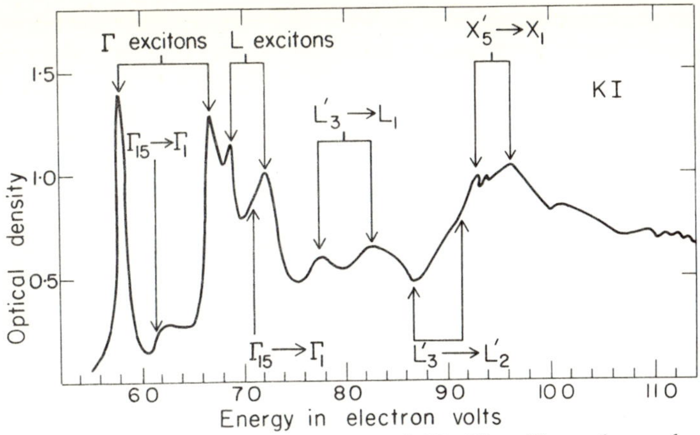

FIG. 8.4. 80°K absorption spectrum of KI (Eby, Teegarden and Dutton.[46] Assignments given by Phillips.[110])

symmetrical antiresonance of the type described earlier in this chapter. The line shapes in the spectrum of annealed Xe have been explained quantitatively by Jain[176] using a generalized random phase approximation which also includes resonant states. So far, symmetrical interferences between resonances have not been seen in the alkali halides but it is highly probable that the strength of these extra excitons is greatly influenced by lattice defects, in the same manner as the band edge excitons. L_3'–L_1 metastable excitons have been identified in NaI, KCl, KBr, KI, RbCl, RbBr and RbI; the energies of the resonances are also shown in Table 8.1. In all except KCl and RbCl the excitons show the characteristic spin–orbit splitting of the parent edge.

8.4. Excitons in layer structure materials

The spectra of strongly anisotropic semiconductors and insulators can provide further evidence of the exciton effects discussed in the previous sections. Very little theoretical work has been done on the band structure of such materials and thus in the examples that follow it is not possible to make definite assignments for the origin of observed resonances. The layer compounds that we consider here (GaSe, GaS, PbI_2 and CdI_2) have hexagonal symmetry within each layer and extremely weak bonding (probably of the van der Waals type) between layers. This bonding causes the easy cleavage perpendicular to the *c*-axis for these materials and is a convenient property from the point of view of sample preparation for either absorption or reflectivity measurements.

TABLE 8.1

Γ and L point exciton levels in the rocksalt structure alkali halides (according to Phillips[110])

The experimental values are taken from the 80°K absorption data of Eby, Teegarden and Dutton [46]

	$\epsilon_\Gamma(3/2)$	$\epsilon_\Gamma(1/2)$	$\epsilon_L(3/2)$	$\epsilon_L(1/2)$
LiF				
LiCl	8·8	8·9		
LiBr	7·20	7·72		
LiI		6·8		
NaF	10·6			
NaCl	7·96	8·09		
NaBr	6·68	7·2		
NaI	5·56	6·73	7·35	8·00
KF	9·95			
KCl	7·76	7·87	9·5	
KBr	6·77	7·26	8·6	8·9
KI	5·80	6·68	6·88	7·25
RbF	9·45			
RbCl	7·51	7·64	8·9	
RbBr	6·60	7·08	7·85	8·15
RbI	5·70	6·47	6·65	6·95
CsF	9·3			

The fundamental spectrum of GaSe reveals a sharp exciton line associated with the lowest direct absorption edge.[158] At low temperature the resonance is resolved into a series of lines. If the series is treated as hydrogenic, and Elliott's theory applied, then a binding energy of 0·037 eV is found. The line width of the first exciton line (0·01 eV), in agreement with cubic materials, is approximately 0·5% of the exciton energy. A note of caution should be added here: in layer structures, the excitons are probably localized to individual layers and have essentially only two-dimensional freedom, thus the model for exciton transitions in isotropic crystals may not apply. One would instead have a series analogous to the two-dimensional hydrogen problem, thus the binding energy quoted above may be too high.

GaS is known to have a very similar band structure to GaSe[158] except that the absorption edge is indirect. The sharp exciton in GaSe is seen only as a very weak peak on the phonon background in GaS. This behaviour (shown in Fig. 8.5), together with the result of a symmetry analysis of GaSe and GaS, suggests that

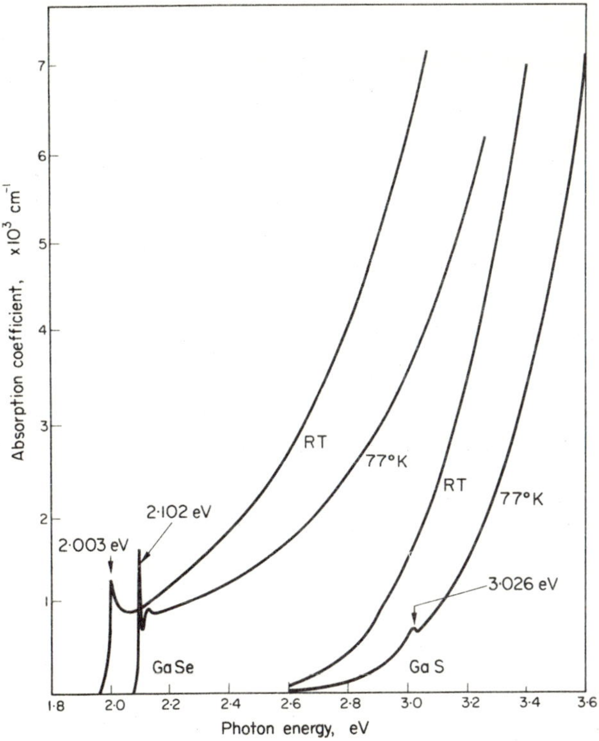

FIG. 8.5. Absorption spectra of GaSe and GaS at 300°K and 77°K ($E \perp c$ axis) (Bassani, Greenaway and Fischer[158]).

the excitons are parabolic and are associated with transitions at the centre of the zone. No antiresonance is seen in GaS because of the relative strength of the scattering continuum. The shape of the exciton assisted edge in GaSe is interesting as it closely resembles the step function behaviour seen in silicon around 3 eV. In Si the line shape is most probably due to hybrid excitons associated with quasi-degenerate M_0 and M_1 edges, but in layer structures a step function in absorption can arise directly from the two-dimensional nature of the crystal lattice, and hence Brillouin zone, as shown in Fig. 4.2. If energy bands in the z-direction of **k**-space (corresponding to the crystal c-axis) are nearly flat for the optical gap states and parabolic for directions such as P or Q (perpendicular to the c-axis), then parabolic bands have essentially cylindrical symmetry and give rise to step functions in the density of states.

PbI_2, a partially ionic layer crystal, provides a good example of both band edge and metastable excitons.[177] The normal incidence reflection spectrum (shown in Fig. 8.6) reveals a wealth

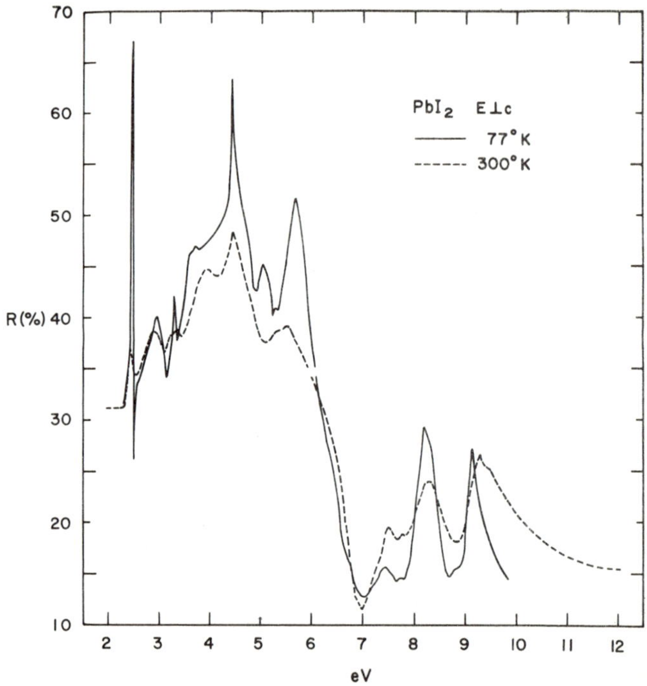

FIG. 8.6. Reflectivity of PbI_2; solid curve 77°K and dashed curve 300°K ($E \perp c$ axis) (Greenaway and Harbeke[179]).

of structure in the ultraviolet which has yet to be analysed. The large changes which take place on cooling clearly indicate that much of the structure is excitonic; in particular the 2·508 eV exciton associated with a direct absorption edge, is followed by an extremely sharp antiresonance caused by interference with the background. Metastable exciton lines occur at 3·25, 4·45 and 5·71 eV. The fact that an absorption edge of some kind is still present in liquid PbI_2[178] indicates that the fundamental gap and its associated exciton are due to direct transitions at the centre of the Brillouin zone. Measurements with linearly polarized light have clearly demonstrated the two-dimensional nature of band edge and metastable excitons in layer compounds.[179]

CdI_2 possesses the same crystal structure as PbI_2. It is an insulator and partially ionic. The fundamental reflectivity spectrum shown in Fig. 8.7 shows evidence of strong interband scattering, and the maxima near 5, 7 and 9 eV do not change

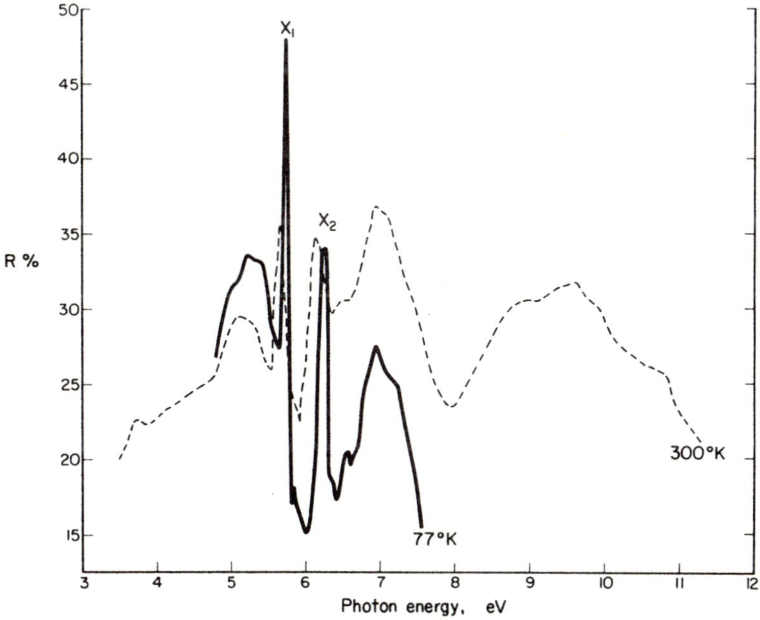

FIG. 8.7. Reflectivity of CdI_2 at 300°K and 77°K showing the formation of metastable excitons (X_1 and X_2 maxima) $E \perp c$ axis (Greenaway and Nitsche[177]).

appreciably on cooling to 77°K. The most remarkable feature is the excitonic structure around 6 eV (X_1 and X_2). Even at room temperature the X_1 and X_2 maxima are separated by a sharp symmetrical antiresonance exactly of the form predicted in Section 8.1. On cooling, the peaks and the antiresonance sharpen further and reveal satellite peaks at higher energies than the main maxima, which are probably the second members of the line series associated with each exciton. In this case the exciton lines fit a three-dimensional hydrogenic series giving binding energies of 0·17 eV and 0·41 eV for X_1 and X_2 respectively, i.e. intermediate between the binding energies for the covalent semiconductors and the alkali halides. The temperature coefficients of X_1 and X_2 are different; this and the different binding energies indicate that the peaks are not coupled together by a mechanism such as spin–orbit splitting but are associated with different states. Nevertheless, it is tempting to draw an analogy with the alkali halides and look to the halogen atoms for the origin of these excitons.

These results on both PbI_2 and CdI_2 show that partially ionic materials can provide interesting examples of the coexistence of strong exciton effects with strong interband effects. Clearly much work remains to be done, particularly from a theoretical point of view, to interpret these spectra. Experimentally, reflectivity measurements (preferably using polarized light so that selection rules may be taken into account) provide an ideal method of investigation. Bulk single-crystal material should be used, and the measurements made at low temperatures in order to observe realistic line shapes.

CHAPTER 9

ELECTRO-OPTICAL EFFECTS. PHOTOELECTRIC EMISSION

9.1. Electroabsorption

The application of an electric field to a crystal can change the symmetry as well as the components of the second-rank tensors n and k describing the optical properties of the crystal. In the past only field-induced changes of n in the long-wavelength transparent region have been investigated and understood as electro-optical effects whereby isotropic crystals become uniaxial and uniaxial crystals become biaxial. The sum rule for the long-wavelength dielectric constant $\epsilon(0) = n_0^2$ in equation (2.37) already shows that the field-induced changes in n_0 must be due to changes in the frequency distribution of interband transitions at and above the energy of the absorption edge. In 1958 Franz[180] and Keldysh[181] calculated the absorption in the presence of a uniform field in the vicinity of the absorption edge by taking electron wave functions containing a linear time dependence of the wave vector. For M_0 type edges the presence of the field results in a modification of the unperturbed square-root behaviour and an additional absorption at energies $\omega < \omega_0$ which is essentially proportional to

$$\exp\left[-\left(\frac{\omega_0 - \omega}{\omega_F}\right)^{3/2}\right] \qquad \text{where} \quad \omega_F^3 = \frac{\epsilon^2 F^2}{2\mu\hbar} \tag{9.1}$$

with the electric field F and the reduced electron-hole mass μ. This gives thus an exponential-type tail in the long-wavelength region with a magnitude increasing exponentially with the applied field. For absorption edges which are already of an exponential character without field, i.e. where the inherent frequency dependence is obscured by other absorption processes, one obtains an edge shift proportional to the square of the applied field or an increase of absorption proportional to exp (F^2). Such edge shifts have been observed in many materials, e.g. in GaAs,[182] where the results agree quantitatively with the shift predicted by Franz's theory taking the proper values of electron and hole mass.

The theoretical treatment has also been extended to energies above the absorption edge by Callaway[183] and Tharmalingam[184] and the results for the absorption in the weak-field limit are expressed in terms of the Airy functions

$$\mathrm{Ai}(\beta) = \frac{1}{\sqrt{\pi}} \int_0^\infty \cos\left(\frac{1}{3}u^2 + u\beta\right)\mathrm{d}u \tag{9.2}$$

as

$$K(\omega,F) = \frac{B}{n\omega}\mu^{3/2}\omega_F{}^{1/2}\left[\left|\frac{\mathrm{dAi}(\beta)}{\mathrm{d}\beta}\right|^2 - \beta\left|\mathrm{Ai}(\beta)\right|^2\right] \tag{9.3}$$

where $\beta = (\omega_0 - \omega)/\omega_F$, ω_F is defined by equation (9.1) and B is a constant containing the dipole matrix element M_{vc} between the band edge states. Due to the oscillatory character of the Airy functions the absorption coefficient oscillates about the curve without field as illustrated in Fig. 9.1(*a*). The period of the oscillations is in practice of the order 5×10^{-2} eV for $F = 10^5$ V/cm. For a fixed frequency it thus gives rise to a periodicity of the absorption as a function of field strength. In addition there is further oscillatory behaviour which has been omitted in equation (9.3) resulting from transitions between the discrete Stark levels produced in the bands by the external field similar to the Landau levels in the presence of a magnetic field[183]. The period of these oscillations is in practice of the order 10^{-3} eV for $F = 10^5$ V/cm.

An expression similar to equation (9.3) has also been derived for the case of indirect transitions.[185–6] First experimental evidence for the oscillatory character of the electroabsorption has come from the measurements in reverse-biased *p–n* junctions in Ge[187–8] and Si,[189–90] where the a.c. field in the junction is near the breakdown field. In the frequency region of indirect transitions the a.c. modulated absorption oscillates around zero and allows a very precise determination of the energies of the phonons involved in the transitions.

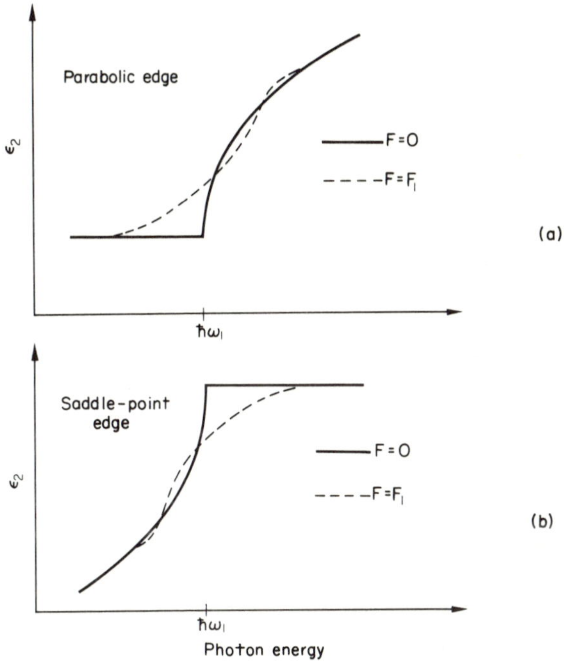

FIG. 9.1. The field-induced change in ϵ_2 plotted as a function of photon energy at (a) a parabolic edge (Callaway[183] and Tharmalingam[184]) and (b) a saddle-point edge (Phillips[195]).

9.2. Electroreflectance

Field-induced changes in the frequency distribution of interband transitions can also be observed in the fundamental reflectivity spectra. In the field-effect arrangement used by Seraphin[191] the field is perpendicular to the reflecting surface and most of the absorption takes place in the space charge region at the surface. The derivative nature of the effect—as in piezoreflectance—enhances the structure due to critical points in the reflectivity spectrum, and thus gives higher resolution. There are also criteria which in addition to the line shape of the direct reflectivity spectra can discriminate among parabolic M_0 edges and M_1 saddle-point edges.

The field dependence of ϵ_2 can be obtained from equation (9.3) simply by using $\epsilon_2 = ncK/\omega$. Subtracting ϵ_2 without field yields $\Delta\epsilon_2$ and the use of the Kramers–Kronig relation in the form

$$\Delta\epsilon_1 = \frac{2}{\pi} \int_0^\infty \frac{\omega' \Delta\epsilon_2(\omega')}{\omega'^2 - \omega^2} d\omega' \tag{9.4}$$

gives the field-induced change in the real part of the dielectric constant. The measured change in reflectivity can be expressed by differentiating Fresnel's formula:

$$\frac{\Delta R}{R} = \alpha(\epsilon_1,\epsilon_2)\Delta\epsilon_1 + \beta(\epsilon_1,\epsilon_2)\Delta\epsilon_2. \tag{9.5}$$

The fractional coefficients α and β depend on ω and can be calculated for each material from the spectra of ϵ_1 and ϵ_2. The curves of α and β for Ge, Si and GaAs[192-3] reveal the dominance of the $\Delta\epsilon_1$ term at low energies. The electroreflectance in the region of the direct absorption edge of Ge has thus to be explained in terms of $\Delta\epsilon_1$.[194] The measurements at and above the fundamental edge of Ge[40] are shown in Fig. 9.2. The curve shows all the transitions at critical points previously determined from the analysis of the direct spectra as narrow peaks or dips. For a detailed analysis of the result we should recall that the theoreti-

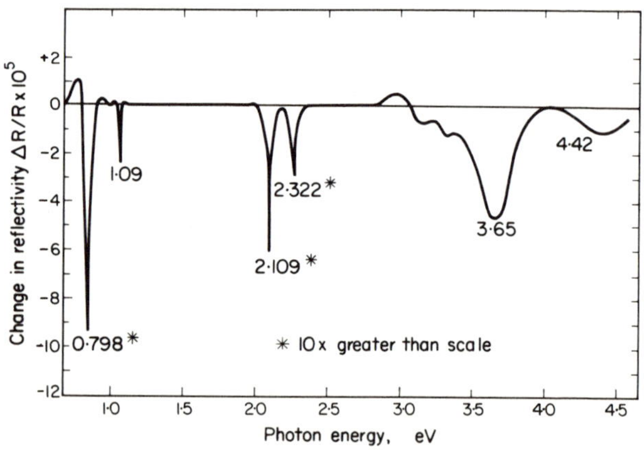

FIG. 9.2. Electroreflectance spectrum of Ge between 0·5 eV and 4·5 eV (Seraphin and Hess[40]).

cal treatment described in Section 9.1 applies only to M_0 type edges whereas most of the ϵ_2 structure is due to saddle points. Phillips,[195] in the so-called duality theorem, has analysed the effect of the electric field on both parabolic and saddle point edges. The analyses for M_1 and M_2 critical points, as Phillips makes clear, gives a correct description only for the longitudinal field direction, i.e. with the field parallel to the direction of the negative reduced mass at an M_1 point or the positive reduced mass at an M_2 point. It is found that the change in ϵ_2 at a saddle-point edge is obtained from the corresponding change at a parabolic edge by reversing the signs of $(\omega-\omega_0)$ and $\Delta\epsilon_2$. This

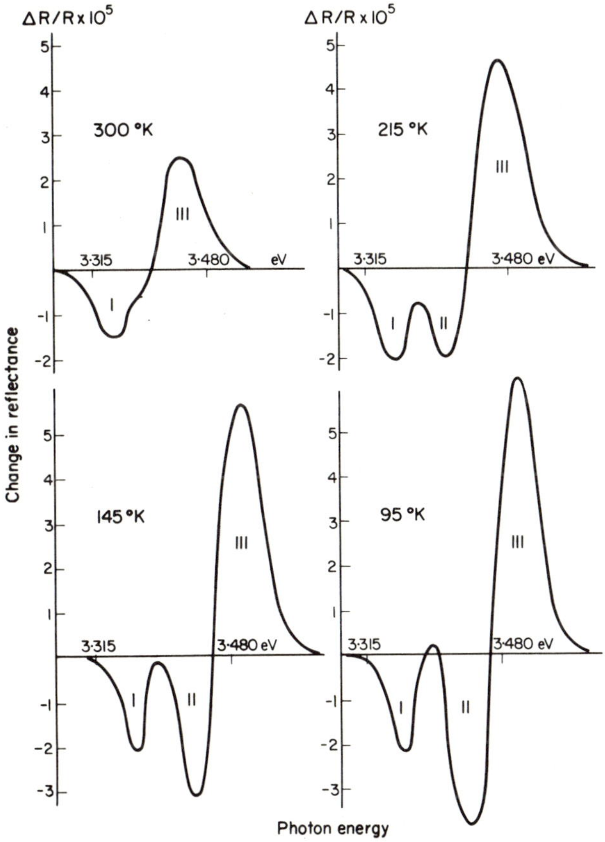

FIG. 9.3. Electroreflectance spectra of Si between 3·3 eV and 3·6 eV at four different temperatures (Seraphin[192]).

behaviour is qualitatively illustrated in Fig. 9.1(b). The important consequence to the line shape of $\Delta R/R$ is that the transformation from a parabolic edge to a saddle-point edge is entirely different for $\Delta\epsilon_1$ and $\Delta\epsilon_2$. ϵ_2 increases with field at ω_0 for a parabolic edge while it decreases for a saddle-point edge. On the other hand, ϵ_1 decreases with field at ω_0 for both types of edge. This has been used to explain the electroreflectance line shape measured in silicon around 3·4 eV.[192] Figure 9.3 shows the results for *p*-type Si at four different temperatures: a complicated structure of maxima and minima is revealed where the direct static reflectivity shows only one maximum. The line II is believed to be associated with exciton formation due to the strong temperature dependence of its size[196]; the rate of autoionization of this metastable exiton is probably enhanced by the applied field. Somewhat different behaviour is observed in *n*-type Si, and Seraphin and Bottka have analysed the 3·4 eV Si results by determining the proper values of α and β which decide the $\Delta\epsilon_1$ and $\Delta\epsilon_2$ contributions, and then choosing the best combination of parabolic or saddle-point edges to fit the experimental curves. It is concluded that for *n*-type material, a parabolic edge at 3·33 eV is followed by a saddle-point edge at 3·41 eV, whereas in *p*-type material two parabolic edges exist at the same energies.[193] The mechanism which can switch the response from one type of edge to its degenerate counterpart is unknown, but may possibly be found in the strength of the potential barrier, which is quite different in the two types of material. As we pointed out, the duality theorem is only valid in restricted cases; Aspnes[197] has recently published the initial results of a generalized theory of electroreflectance, and the assignments that we have discussed above must still be regarded as tentative until they can be confirmed by a more generalized treatment of the effect.

The ability of the electroreflectance method to resolve fine structure which is obscured in direct optical spectra has also been demonstrated for a number of zincblende materials. Figure 9.4 shows the room temperature electroreflectance spectra of some III V compounds in the region of the E_0 peak, obtained by Shaklee, Cardona and Pollak,[149] who applied the electric field at a semiconductor–electrolyte interface. The E_0 peak previously attributed to Γ_{15}–Γ_{15} transitions is very weak in the ϵ_2 spectra of these materials. In the electroreflectance they are very pronounced and exhibit a doublet nature. It appears that this splitting Δ_0' is caused by spin–orbit interaction since it increases with average atomic number in the same way as the splitting of Γ_{25}' and Λ_3. Since the values are smaller than either of the latter two splittings it is believed that Δ_0' is due to a splitting of the top valence band in the ⟨100⟩ direction. This is in quantitative agreement with the results of band structure calculations with the **k·p**

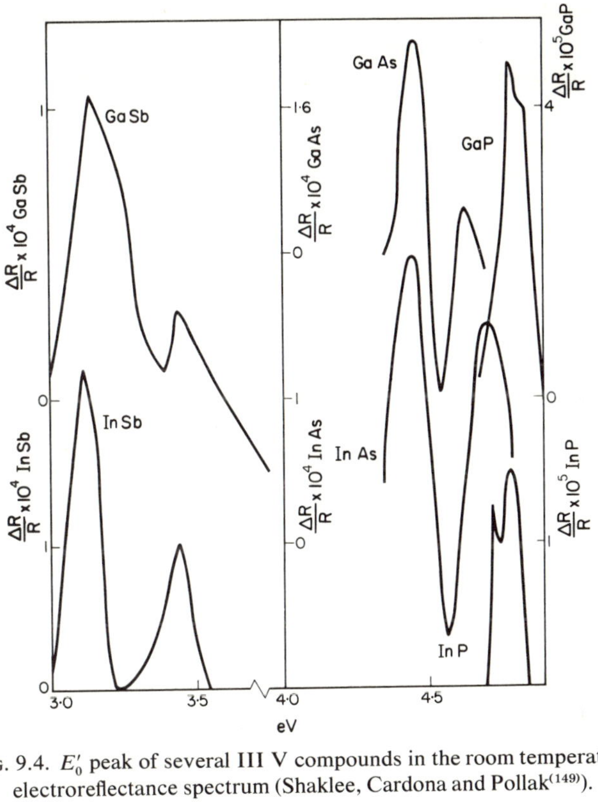

FIG. 9.4. E_0' peak of several III V compounds in the room temperature electroreflectance spectrum (Shaklee, Cardona and Pollak[149]).

technique including the spin–orbit interaction (as shown in Fig. 5.3). These results also show that the Δ_5 valence bands and the Δ_3 conduction band are nearly flat near $\mathbf{k} = 0$. The experiments thus give evidence that the E_0' peaks are caused by transitions between these bands in the ⟨100⟩ direction not far from Γ.

In ending this section we must realize that the first electroreflectance results were obtained as recently as 1964. As with most new techniques the wealth of experimental data at present exceeds the tools available for their full analysis. The bulk of the current data has served to provide enhanced resolution over static measurements, and has permitted more detailed studies to be made of the effects of, for example, temperature, spin–orbit interaction, and lifetime broadening. The higher resolution of this differential method has so far added little definitive new information to that from the previous static measurements of the optical constants. One of the main obstacles is still the analysis of the line shapes obtained in electroreflectance which may be strongly influenced by exciton-quenching effects.[198] It is clear, however, that with the advent of a detailed theoretical treatment, an extremely powerful tool for the study of the band structure of solids will be available.

9.3. Photoelectric emission

The study of photoelectric emission from solids has been widely used for the determination of the band structure as a method complementary to the reflectivity spectra analysis. Only this aspect of photoemission will be discussed here. The technique provides an additional parameter for the band structure analysis, since besides the spectral dependence of the total photoelectric yield, the energy distribution of the emitted electrons for fixed incident photon energy gives valuable information about the position of valence and conduction band states separately.

Maxima in the yield curve corresponding to ϵ_2 peaks arise if the final state involved in the primary optical transition lies above the vacuum level, and minima occur if the final state lies below the vacuum level, since the photons are then absorbed without giving rise to emission. The structure is enhanced by the fact that only electrons very near to the surface will be able to leave the crystal. The yield will therefore be large if the light is absorbed close to the surface, i.e. in spectral regions of high absorption coefficient caused by critical points, provided the escape probability is not energy dependent.

The energy of the emitted electrons is equal to the energy of the final state in the conduction band with respect to the vacuum level. This is very important for the determination of the band positions since the optical spectra yield only energy differences between valence and conduction band states and the joint density of states.

An inherent disadvantage of the photoemissive process is that only electrons above the vacuum level (which in pure crystals is several electronvolts higher than the Fermi level) can be studied. This difficulty has been overcome by depositing a thin layer of caesium on the surface whereby the electron affinity of, for example, Si is reduced from about 4 eV to 1·5 eV.[199] However, it has been shown that 0·3 of a monolayer of Cs on a germanium surface results in a non-conservation of the $\mathbf{k}_{\parallel}$ component tangential to the surface whereas $\mathbf{k}_{\parallel}$ is conserved for a significant fraction of the electrons in the emission process from an ultra-clean surface.[200] Such dynamic effects occurring during the escape process give rise to changes and broadening of the energy distribution of the emitted electrons.

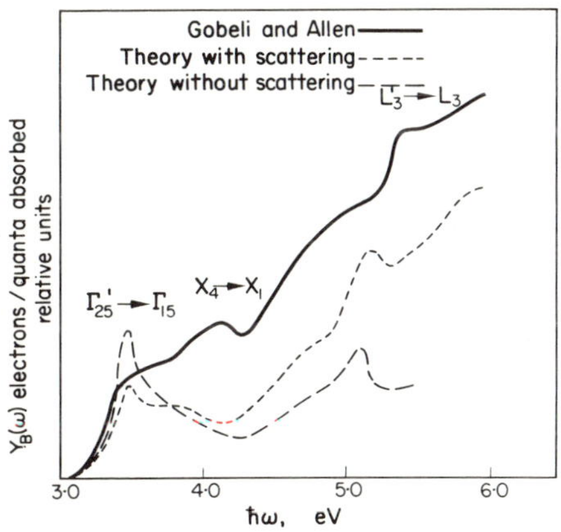

FIG. 9.5. Spectral dependence of the photoemissive yield of Si.[146] Solid line: experimental data of Gobeli and Allen. Dashed lines: theoretical curves of Brust.

It has, however, been possible to explain the yield curves of pure and caesiated Si, Ge and several III V compounds,[42,201] in terms of the band structure with an arbitrary wavelength-independent escape probability for electrons above the vacuum level,[73,202] or with an escape probability arising from a complete randomization of the crystal momentum due to scattering before the electron reaches the surface.[146] Figure 9.5 shows as an example the experimental yield $Y_B(\omega)$ of Si with nearly one monolayer of Cs according to Allen and Gobeli, and the calculated curves of Brust where the vacuum level was taken at 2·5 eV. It is noted that the theoretical curve for the case of total scattering is closer to the experiment as expected from the presence of the deposited layer. The experimental curve carries the assignment of maxima or minima such as X_4–X_1 where the final X_1 state is well below the vacuum level, thus giving rise to a dip in the yield curve. In this case the energy distribution has also been calculated from the density of states in valence and conduction bands but the results can only explain the basic features of the data; this is probably due to the neglect of dynamic effects. Kane[(203)] has pointed out that pronounced structure in the photoelectric energy distribution curves may result from transport effects such as low group velocity and enhanced phonon scattering occurring at a peak in the density of states.

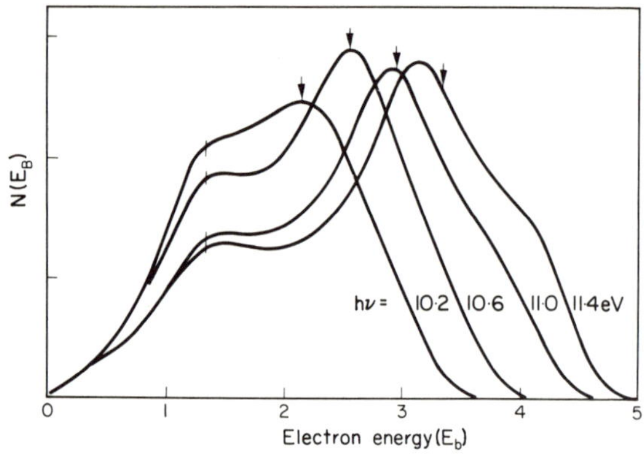

FIG. 9.6. Energy distribution of electrons emitted from CdS cleaved in high vacuum for four different incident photon energies $h\nu$ (Kindig and Spicer[(204)]).

Measurements on CdS have been performed on uncaesiated surfaces which have been cleaved either in high or low vacuum.[(204–5)] Figure 9.6 shows typical energy distributions of a crystal cleaved in high vacuum for four different photon energies. We note one maximum at 1·35 eV independent of photon energy which is more pronounced for smaller incident photon energies. The position of the other maximum varies with photon energy as

$E_{max} = (\hbar\omega - 8{\cdot}05)$ eV. Maxima of the first kind indicate regions of high density of states in the conduction band, those of the second kind are due to such regions in the valence band. For direct optical transitions and emission without scattering, the electrons should be emitted with energy $E = \hbar\omega - E_G - E_v - W$, where E_G is the energy gap, E_v the energy of the initial state relative to the valence band maximum, and W the electron affinity, i.e. the energy of the vacuum level with respect to the conduction band minimum.

The maxima in the energy position should shift with increasing $\hbar\omega$ somewhat less than $\hbar\omega$ itself due to the variation of E_v. In order to explain the actual $\hbar\omega$ dependence of the maximum positions the concept of non-direct transitions, i.e. non-conservation of **k** in the absorption process, has been invoked.[43, 205–6] As shown by the temperature dependence, these are not phonon-assisted transitions. This concept is difficult to reconcile with the optical data on CdS at least in the low-energy region and particularly where the strong polarization effects reported for wurtzite-type materials require **k** conservation.[120] It should be noted, however, that the photoemission results obtained from uncaesiated surfaces concern only electrons in high conduction band states where the electron–electron interaction is stronger than at lower energies thus giving rise to Auger excitations. On the other hand, most of the structure in $\epsilon_2(\omega)$ is associated with states in the lower conduction bands. It should be noted that the non-conservation of **k** is only claimed for tightly bound valence bands, i.e. if the hole produced by the excitation is localized within a unit cell.[207] This is a situation in which no full translational invariance of the lattice is provided, so that the band concept for the localized hole breaks down. For this reason, it may thus happen that in a given material, transitions from *s*- or *p*-like valence bands are **k**-conserving, whereas transitions from localized *d*-states are not. The question of direct versus indirect transitions has also been discussed in detail in a photoemission study of the band structure of PbTe,[206] where in conjunction with the earlier ϵ_2 data[10] a reasonable agreement with the theoretical band structure calculation[102] has been established.

CHAPTER 10

PLASMA AND ELECTRON ENERGY-LOSS PHENOMENA

10.1. D-band effects

We have seen that the fundamental optical spectra of materials such as the III V zincblende semiconductors are dominated by strong interband transitions occurring between the absorption edge and about 8 eV. Around 20 eV further weak structure is observed due to the onset of real transitions from atomic *d*-electron levels into the conduction band. Between these two regions the optical spectra are characterized by a rapid and smooth decrease in the reflectivity or absorption which is similar to the behaviour seen in certain metals in the ultraviolet. In this so-called plasma region the valence electrons can behave like free particles and take part in collective oscillations. The reflectivity curve for InSb illustrates this behaviour well, and is shown in Fig. 10.1.

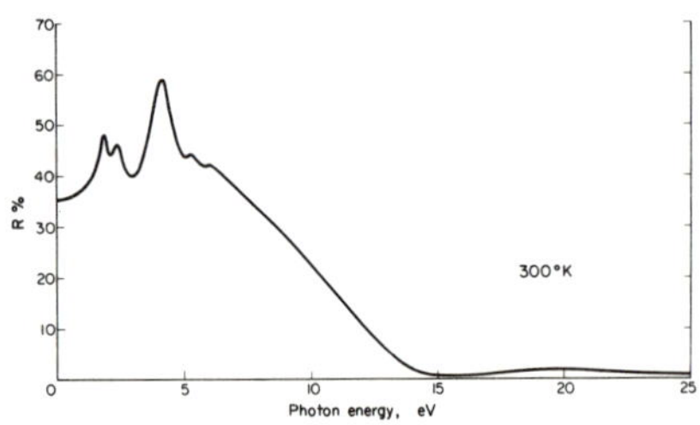

FIG. 10.1. Room temperature reflectance of InSb (Ehrenreich and Philipp[85]).

In a many electron system the transverse and longitudinal dielectric constants determine, respectively, the optical and the collective properties. At long wavelengths in the random phase approximation these dielectric constants are equal. For cubic insulating or semiconducting crystals, in the long-wavelength limit which is relevant in the fundamental absorption region, Ehrenreich and Cohen[50] have shown that the frequency-dependent dielectric constant $\epsilon(\omega)$ can be written in the form of equation (4.8). One of the limiting cases of equation (4.8) concerns the situation where a single group of valence bands v is present which is well separated from the core states. Then at frequencies sufficiently high so that the oscillator strengths of transitions from v to higher bands are exhausted, the dielectric constant can be written:

$$\epsilon(\omega) \simeq 1 - \frac{\omega_{pv}^2}{(\omega + i/\tau_v)^2}, \tag{10.1}$$

which is equivalent to equation (2.33) with a complex frequency. As used in equation (2.35) $\omega_{pv}^2 = N_v e^2/m\epsilon_0$, where N_v is the density of valence electrons. The relaxation time now depends only on the band index v. This limiting case is essentially a modified Drude formula involving the free electron mass m and indicates that the valence electrons behave like free particles.

The situation in equation (10.1) is realized in practice in Si where the valence band lies approximately 100 eV above the next highest $2p$ core states. Thus in Si above about 6 eV the behaviour of $\epsilon(\omega)$ and hence the optical constants are described extremely well using equation (10.1) and a value of N_v of 4 electrons per atom.

When d-bands are present, as in Ge and the zincblende semiconductors, equation (10.1) is modified by the oscillator strength coupling between the valence and d-band electrons and can now be written:

$$\epsilon(\omega) \simeq \left(1 + \delta\epsilon(0)\right)\left[1 - \frac{\Omega_{pv}^2}{(\omega + i/\tau_v)^2}\right]. \tag{10.2}$$

Here $\delta\epsilon(0)$ corresponds to the contribution of the d-electrons to the static dielectric constant $\epsilon(0)$ and Ω_{pv} is an effective plasma frequency which is modified from the free electron value by the presence of the d-bands. Methods of estimating both $\delta\epsilon(0)$ and Ω_{pv} from optical data by means of the sum rules derived in Section 2.4 have been described by Ehrenreich and Philipp.[85]

In Fig. 10.2 a plot of n_{eff} defined by equation (2.36) versus energy for a series of semiconductors is shown. It is seen that

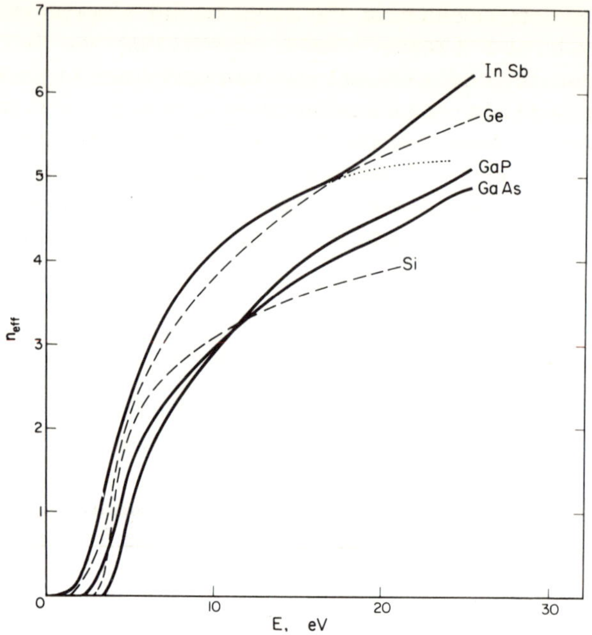

FIG. 10.2. n_{eff} plotted against photon energy from group IV and III V semiconductors (Ehrenreich and Philipp[85]).

for silicon with no d-bands, n_{eff} saturates at a value of approximately 4; for the other materials n_{eff} rises smoothly to a value well in excess of 4 because of the valence d-band coupling. Beyond 20 eV a further abrupt rise takes place due to the onset of real transitions from the d-band into the conduction band. This picture is strongly supported by the electron energy loss spectra of Ga and In where the excitation of d-electrons is seen around 20 eV.[208] An extrapolation of n_{eff} to higher energy, before the onset of the real d-conduction band transitions, enables Ω_{pv} to be estimated.

The quantity $\epsilon(0)_{eff}$ defined by equation (2.38) is plotted as a function of energy in Fig. 10.3 for the same materials which were shown in Fig. 10.2. It is seen immediately that beyond about 10 eV the curve for Si saturates at a value $\epsilon(0)_{eff} = \epsilon(0)$, i.e. there exist no further contributions to $\epsilon(0)$ from higher energy transitions. For the other materials the contribution to $\epsilon(0)$ caused by the d-electrons $\delta\epsilon(0)$ is just the difference between $\epsilon(0)_{eff}$ and the static dielectric constant above the region where valence-conduction band transitions take place (the sharp rise in Fig. 10.3) but below the region where real transitions from *d*-bands into the conduction band can take place.

Using the values of the modified plasma frequency Ω_{pv} and $\delta\epsilon(0)$ equation (10.2) can be used to calculate the behaviour of ϵ_1 and ϵ_2 in the plasma region. The only unknown parameter is the relaxation time τ_v which is obtained by fitting equation (10.2) with experimental values of ϵ_1 and ϵ_2 obtained from optical data

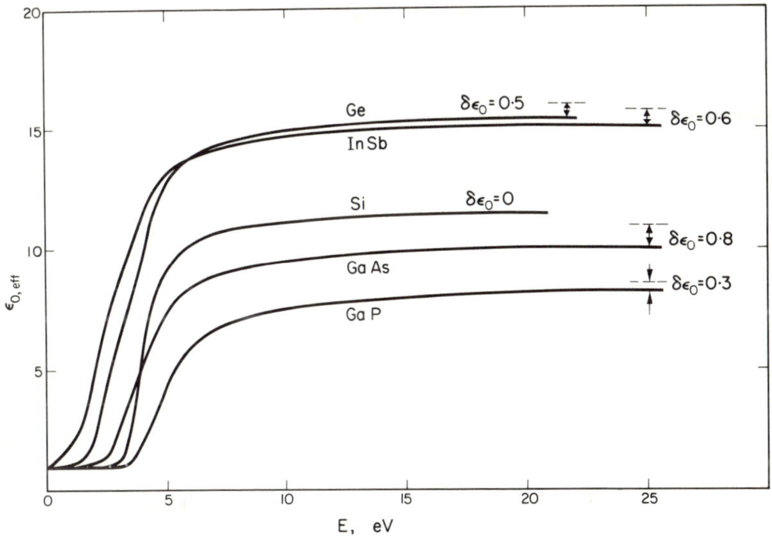

FIG. 10.3. $\epsilon(0)_{\text{eff}}$ plotted against photon energy for some group IV and III V semiconductors (Ehrenreich and Philipp[85]).

via the Kramers–Kronig relations. For the materials shown in Fig. 10·3 τ_v lies in the range $1{\cdot}4 \times 10^{-16}$–$1{\cdot}8 \times 10^{-16}$ secs. The magnitude of τ_v indicates that electron–electron scattering is the dominant broadening mechanism in the plasma region. Lifetimes due to lattice or impurity scattering are expected to be several orders of magnitude greater than this.

Very similar values have been obtained for the scattering times in SnTe.[10] If one plots n_{eff} versus energy according to equation (2.36) for this material[95] (see Fig. 10.4) one finds a change in slope at 6 eV. At 20 eV n_{eff} is almost one electron per atom higher than the value extrapolated from lower energies, indicating that one additional electron per atom contributes to the optical properties above 6 eV. Similar results are found for the other materials of the lead salt family. This confirms the view taken in Section 5.6 that the additional ϵ_2 structure in this region is due to transitions from the second lowest Γ_1 state formed by $6s^2$ electrons from Pb or Sn to the conduction band states.

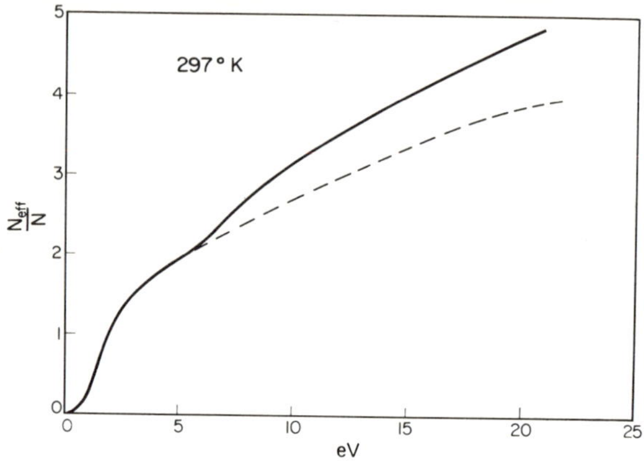

FIG. 10.4. Effective number of electrons per atom plotted against photon energy for SnTe (Cardona[95]).

The curve for n_{eff} in graphite[134] evaluated from ϵ_2 data shows a plateau at nearly 1 electron per atom between 6 and 10 eV and rises abruptly at higher energies to the saturation value of 4. This gives clear evidence that all optical transitions up to 10 eV occur between π-bands formed by one p-electron per atom and that the onset of σ–σ transitions is above 10 eV as discussed in Section 6.2.4. A calculation of n_{eff} from the optical data can thus provide valuable information as to the origin of the optical transitions.

10.2. Energy-loss spectra

The equivalence of the transverse and longitudinal dielectric constants in the frequency range under consideration allows us to compare the loss spectrum calculated from optical data with the energy loss spectrum of fast electrons traversing a crystal. The loss function $-\text{Im}\,(1/\epsilon) = \epsilon_2/(\epsilon_1^2 + \epsilon_2^2)$ involves both one electron excitations (interband transitions) and many body resonances such as excitons or plasmons. This is illustrated in Fig. 10.5 where the energy loss spectra of KCl and KBr obtained by both techniques are compared. The optical data[109] (dashed line) have been multiplied by a function which takes the angular and energy distribution of the primary electron beam into account, and normalized in order to make both curves comparable. The striking agreement in energies and line shape proves the equivalence of the transverse and longitudinal dielectric constant and confirms the validity of the random phase approximation in this

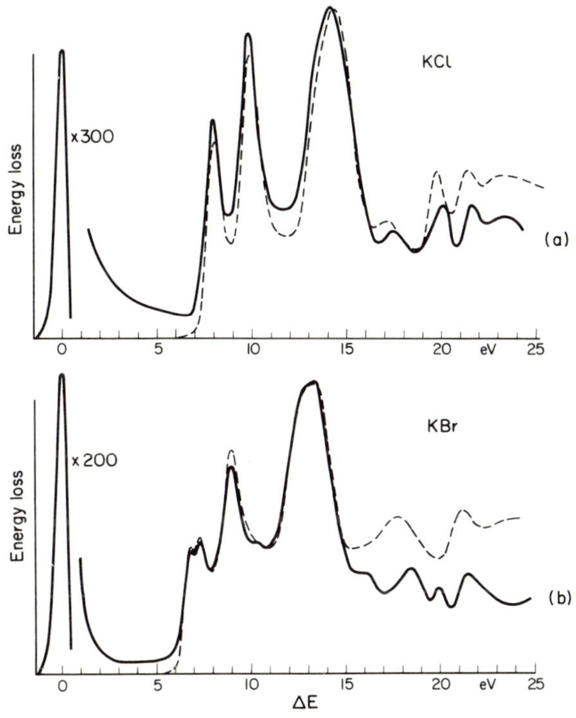

FIG. 10.5. Energy loss spectra of KCl and KBr. The solid lines refer to direct experimental data obtained by Creuzburg and Raether[209]. Dashed lines represent the loss function calculated from optical data.

energy range. Most of the low-energy structure is due to excitation of excitons, the electron energy loss curve[208–9] even reproduces clearly the spin–orbit splitting of the exciton line associated with the $\Gamma_{15}-\Gamma_1$ transitions in KBr at room temperature. The main plasma resonances in KCl and KBr occur at 13·9 and 13·4 eV respectively. As in many other materials, the condition for a plasma resonance that $\epsilon_1(\omega) = 0$ in a region where the complex $\epsilon(\omega) = 0$ is not rigidly fulfilled. Ehrenreich and Philipp have pointed out[85] that the maximum of Im $(1/\epsilon)$ will correspond to a plasma resonance if (i) ϵ_1 and ϵ_2 are small in the region of the peak, (ii) ϵ_1 and ϵ_2 vary approximately linearly with $d\epsilon_1/d\omega > 0$ and $d\epsilon_2/d\omega < 0$, and (iii) $(\omega/\epsilon_1)\ d\epsilon_1/d\omega \gg 1$ and $(\omega/\epsilon_2)(d\epsilon_2/d\omega) \gg 1$. These rules are approximately obeyed in the region of the 13.4 loss peak in KBr, whereas the calculated value taking eight electrons per molecule is 12·3 eV. Plasma frequencies are shifted by neighbouring interband and exciton transitions to higher energies if the transitions lie below the unperturbed plasma frequency and vice versa. The shift of about 1 eV to higher energies in KBr results from the strong exciton structure at lower energies.

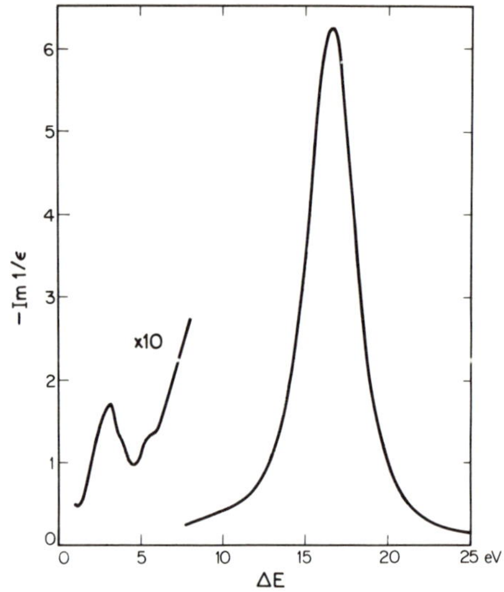

FIG. 10.6. Energy loss spectrum of Si (Zeppenfeld and Raether[209]).

In the group IV and III V semiconductors the plasma region can be readily recognized from primary data such as reflectivity curves by the long structureless decrease in R typified by the curve for InSb. In these materials the plasma resonance corresponds to $\epsilon_1(\omega) \simeq 0$. The dominating character of the main plasma resonance in the loss spectra is typified by the curve for Si shown in Fig. 10.6.[210] The plasma frequency is in agreement with the calculated value as well as that derived from the optical data. The half-width of the peak (3·6 eV) is, however, only half of that of the optical loss curve, and the same holds for measurements on Ge. This illustrates that the energy loss data are more accurate in the plasma region where ϵ_1 and ϵ_2 are very small if the data are not obscured by surface losses or multiple losses. As outlined in Section 2.3 the dielectric constant can also be deduced from energy loss data with the Kramers–Kronig relations as demonstrated by LaVilla and Mendlowitz[(211)] for Al. This procedure has not yet been followed for semiconductors but it should provide accurate values for ϵ_1 and ϵ_2 in this region.

At lower energies, on the other hand, the optical data are superior since the ϵ_2 structure is reduced by the factor $(\epsilon_1^2+\epsilon_2^2)^{-1}$ in the energy-loss spectra. Only in films which are thicker than normally used in these experiments does the structure in ϵ_2 become detectable (see, for example, the peak at 2·9 eV in Fig. 10.6). Both methods therefore should be used in their appropriate frequency regions as an overall check on the consistency of the results.

REFERENCES

1. Stern, F. Elementary theory of the optical properties of solids, *Solid State Physics* **15**, 299–408 (1963).
2. Moss, T. S. *Optical Properties of Semiconductors*, Butterworth, 1959.
3. Stratton, J. A. *Electromagnetic Theory*, McGraw-Hill, 1941.
4. Toll, J. S. *Phys. Rev.* **104**, 1760–70 (1956).
5. Landau, L. D. and Lifshitz, E. M. *Electrodynamics of Continuous Media*, Pergamon, 1960.
6. Kronig, R. de L. *J. Opt. Soc. Am.* **12**, 547–57 (1926).
7. Bode, H. W. *Network Analysis and Feedback Amplifier Design*, Van Nostrand, 1945.
8. Roessler, D. M. *Brit. J. Appl. Phys.* **16**, 1359–66 (1965).
9. Velický, B. *Czech. J. Phys.* B, **11**, 787–98 (1961).
10. Cardona, M. and Greenaway, D. L. *Phys. Rev.* **133**A, 1685–97 (1964).
11. Philipp, H. R. and Taft, E. A. *Phys. Rev.* **113**, 1002–5 (1959).
12. Philipp, H. R. and Ehrenreich, H. *Phys. Rev.* **129**, 1550–60 (1963).
13. Hass, G., Hunter, W. R. and Tousey, R. *J. Opt. Soc. Am.*, **46**, 1009–12 (1956).
14. Angel, D.W., Hunter, W.R., Hass, G. and Tousey R. *J. Opt. Soc. Am.*, **51**, 913–4 (1961).
15. Hass, G. and Tousey, R. *J. Opt. Soc. Am.* **49**, 593–602 (1959).
16. Seya, M. *Sci. of Light (Tokyo)*, **2**, 8–17 (1952).
17. Namioka, T. *Sci. of Light (Tokyo)*, **3**, 15–24 (1954).
18. Namioka, T. *J. Opt. Soc. Am.* **49**, 951–61 (1959).
19. Huffman, R. E., Hunt, W. W., Jr., Tanaka, Y. and Novack, R. L. *J. Opt. Soc. Am.* **51**, 693 (1961).
20. Hopfield, J. J. *Astrophys. J.* **72**, 139 (1930).
21. Tanaka, Y., Jursa, A. S. and LeBlanc, F. J. *J. Opt. Soc. Am.* **48**, 304–8 (1958).
22. Tanaka, Y. and Zelikoff, M., *J. Opt. Soc. Am.* **44**, 254–5 (1954).
23. Hunter, W. R., *Proc. 10th Int. Colloq. on Spectrosc.*, College Park, Maryland, 1962.
24. Tousey, R. *Appl. Optics* **1**, 679–94 (1962).
25. Dunkelman, L., Fowler, W. B. and Hennes, J. *Appl. Optics* **1**, 695–700 (1962).
26. Childs, C. B. *J. Opt. Soc. Am.* **51**, 895–7 (1961); Hinteregger, H. E. and Watanabe, K., *ibid.* **43**, 604–8 (1953).
27. Baker, R. F. *J. Opt. Soc. Am.* **28**, 55–60 (1938).
28. Hass, G., and Hunter, W. R., *J. Quant. Spectrosc. and Radiative Transfer (G.B.)* **2**, 637–46 (1962).
29. Elliott, A., Ambrose, E. J. and Temple, R. *J. Opt. Soc. Am.* **38**, 212–16 (1948).
30. Robin, M. B., Kuebler, W. A. and Yoh-Han Pao, *Rev. Sci. Instr.* **37**, 922–4 (1966).

31. HAMM, R. N., MACRAE, R. A. and ARAKAWA, E. T. *J. Opt. Soc. Am.* **55**, 1460–3 (1965).
32. FAN, H. Y. *Methods of Experimental Physics* **6B**, 258–62 (1959).
33. SCOULER, W. J. *Appl. Optics* **3**, 341–3 (1964).
34. NEURINGER, L. J., *Phys. Rev.* **113**, 1495–1503 (1959).
35. MICHELS, A., VAN ECK, J., MACHLUP, S. and TEN SELDAM, C. A. *J. Phys. Chem. Solids* **10**, 12–18 (1959).
36. GERHARDT, U. *Phys. Stat. Sol.* **11**, 801–18 (1965).
37. PHILIPP, H. R., DASH, W. and EHRENREICH, H. *Phys. Rev.* **127**, 762–5 (1962).
38. ENGELER, W. E., FRITZSCHE, H., GARFINKEL, M. and TIEMANN, J. J., *Phys. Rev. Letters* **14**, 1069–72 (1965).
39. GOBELI, G. W. and KANE, E.O. *Phys. Rev. Letters* **15**, 142–6 (1965).
40. SERAPHIN, B. O. and HESS, R. B., *Phys. Rev. Letters* **14**, 138–40 (1965).
41. SHAKLEE, K. L., CARDONA, M. and POLLAK, F. H. *Phys. Rev. Letters* **15**, 883–5 (1965).
42. SPICER, W. E. and SIMON, R. E., *Phys. Rev. Letters* **9**, 385–9 (1962).
43. SPICER, W. E. and KINDIG, N. B., *Solid State Communications* **2**, 13–16 (1964).
44. SOMMER, A. H. and SPICER, W. E., *Methods of Experimental Physics* **6B**, 376–91 (1959).
45. HOLMES, P. J. (Editor) *Electrochemistry of Semiconductors* (Academic Press, 1962), and GATOS, M. C. and LAVINE, M. C. Chemical behaviour of semiconductors (etching characteristics), *Progress in Semiconductors* **9**, 1–45 (1965).
46. EBY, J. E., TEEGARDEN, K. J. and DUTTON, D. B. *Phys. Rev.* **116**, 1099–1105 (1959).
47. GAVINI, A., SUSSMANN, R. and CARDONA, M. *Bull. Am. Phys. Soc.* **11**, 85 (1966).
48. GREENAWAY, D. L. and HARBEKE, G. *J. Phys. Chem. Solids* **26**, 1585–1604 (1965).
49. BASSANI, F. *Band Structure and Interband Transitions*, Rendiconti della Scuola Internationale "E. Fermi" Corso XXXIV, Academic Press, 1966.
50. EHRENREICH, H. and COHEN, M. *Phys. Rev.* **115**, 786–90 (1959).
51. VAN HOVE, L. *Phys. Rev.* **89**, 1189–93 (1953).
52. PHILLIPS, J. C. *Phys. Rev.* **104**, 1263–77 (1956).
53. MCLEAN, T. P. The absorption edge spectrum of semiconductors, *Progress in Semiconductors* **5**, 53–102 (1960).
54. CARDONA, M., *Phys. Rev.* **129**, 69–78 (1963).
55. JONES, H. *The Theory of Brillouin Zones and Electronic States in Crystals*, North-Holland, 1960.
56. HEINE, V. *Group Theory in Quantum Mechanics*, Pergamon, 1960.
57. TINKHAM, M. *Group Theory and Quantum Mechanics*, McGraw-Hill, 1964.
58. HAMERMESH, M. *Group Theory and its Application to Physical Problems*, Pergamon, 1962.
59. BOUCKAERT, L. P., SMOLUCHOWSKI, R. and WIGNER, E. P. *Phys. Rev.* **50**, 58–67 (1936).
60. KOSTER, G. F., DIMMOCK, J. O., WHEELER, R. G. and STATZ, H. *Properties of the Thirty-two Point Groups*, M.I.T. Press, 1963.
61. CALLAWAY, J. *Energy Band Theory*, Academic Press, 1964.
62. HARRISON, W. A. *Phys. Rev.* **118**, 1190–1208 (1960).
63. HERRING, C. *Phys. Rev.* **57**, 1169–77 (1940).
64. HERMAN, F. *Phys. Rev.* **93**, 1214–25 (1954).

65. PHILLIPS, J. C. *Phys. Rev.* **112**, 685–95 (1958).
66. BRUST, D., PHILLIPS, J. C. and BASSANI, G. F. *Phys. Rev., Letters* **9**, 94–7 (1962).
67. SEITZ, F. *The Modern Theory of Solids*, McGraw-Hill, 1940.
68. MARPLE, D. T. F. and EHRENREICH, H. *Phys. Rev. Letters* **8**, 87–9 (1962).
69. PHILLIPS, J. C. The fundamental optical spectra of solids, *Solid State Physics* **18**, 55–164 (1966).
70. TAUC, J. and ANTONČIK, E., *Phys. Rev. Letters* **5**, 253–4 (1960).
71. PHILIPP, H. R. and TAFT, E. A. *Phys. Rev.* **120**, 37–8 (1960).
72. TAUC, J. and ABRAHAM, A. *J. Phys. Chem. Solids* **20**, 190–2 (1961).
73. BRUST, D., COHEN, M. L. and PHILLIPS, J. C. *Phys. Rev. Letters* **9**, 389–92 (1962).
74. BRUST, D. *Phys. Rev.* **134A**, 1337–53 (1964).
75. POTTER, R. F. *Proc. Int. Conf. on Semiconductor Physics,* Kyoto, 1966, pp. 107–10.
76. KANE, E. O. *Phys. Rev.* **146**, 558–67 (1966).
77. HERMAN, F., KORTUM, R. L., KUGLIN, C. D. and SHORT, R. A. *Quantum Theory of Atoms, Molecules, and the Solid State*, P. O. Löwdin (Ed.), Academic Press, 1966.
78. PHILIPP, H. R. and TAFT, E. A. *Phys. Rev.* **127**, 159–61 (1962).
79. CARDONA, M. and GREENAWAY, D. L. *Phys. Rev.* **125**, 1291–6 (1962).
80. GROVES, S. and PAUL, W. *Phys. Rev. Letters* **11**, 194–6 (1963).
81. HERMAN, F. *Phys. Rev.* **88**, 1210–11 (1952).
82. BASSANI, F. and LIU, L. *Phys. Rev.* **132**, 2047–50 (1963).
83. SASLOW, W., BERGSTRESSER, T. K. and COHEN, M. L. *Phys. Rev. Letters* **16**, 354–6 (1966).
84. CARDONA, M. and POLLAK, F. H. *Phys. Rev.* **142**, 530–43 (1966).
85. EHRENREICH, H. and PHILIPP, H. R. *Proc. Int. Conf. on Semiconductor Physics*, Exeter, 1962, pp. 367–74.
86. TAUC, J. and ABRAHAM, A., *Proc. Int. Conf. on Semiconductor Physics,* Prague, 1960, pp. 375–9.
87. GREENAWAY, D. L. *Phys. Rev. Letters* **9**, 97–8 (1962).
88. PAUL, W., *J. Appl. Phys.* **32**, 2082–94 (1961).
89. CARDONA, M. and GREENAWAY, D. L. *Phys. Rev.* **131**, 98–103 (1963).
90. CARDONA, M. *Z. Physik* **161**, 99–102 (1961).
91. ZALLEN, R. and PAUL, W. *Phys. Rev.* **134**A, 1628–41 (1964).
92. THOMPSON, A. G., CARDONA, M., SHAKLEE, K. L. and WOOLLEY, J. C. *Phys. Rev.* **146**, 601–10 (1966).
93. SCOULER, W. J. and WRIGHT, G. B. *Phys. Rev.* **133**A, 736–9 (1964).
94. PHILLIPS, J. C. *Phys. Rev.* **133**A, 452–9 (1964).
95. CARDONA, M., *J. Appl. Phys.* **36**, 2181–6 (1965).
96. COHEN, M. L. and BERGSTRESSER, T. K. *Phys. Rev.* **141**A, 789–96 (1966).
97. BASSANI, F., KNOX, R. S. and FOWLER, W. B. *Phys. Rev.* **137**A, 1217–25 (1965).
98. SCOP, P. M. *Phys. Rev.* **139**A, 934–40 (1965).
99. HERMAN, F. *J. Electronics* **1**, 103–14 (1955).
100. BASSANI, F. and CELLI, V. *J. Phys. Chem. Sol.* **20**, 64–75 (1961).
101. CARDONA, M. *Proc. Int. Conf. on Semiconductor Physics*, Paris, 1964, pp. 181–96.
102. See references 1–24 in LIN, P. J. and KLEINMAN, L. *Phys. Rev.* **142**, 478–89 (1966).
103. GIBSON, A. F. *Proc. Phys. Soc. (London)* **B63**, 756–67 (1950); **B65**, 378–88 (1952).

104. JOHNSON, L. E., SWITENDICK, A. C., CONKLIN, J. B. and PRATT, G. W., Jr. *MIT Solid State and Molecular Theory Group,* Progress Report No. 46 (1962).
105. JOHNSON, L. E., CONKLIN, J. B. and PRATT, G. W., Jr. *Phys. Rev. Letters* **11**, 538–40 (1963).
106. HALL, J. J. and KOENIG, S. H. *Bull. Am. Phys. Soc.* **8**, 51 (1963).
107. FALICOV, L. M. and LIN, P. J. *Phys. Rev.* **141**, 562–7 (1966); LIN, P. J. and PHILLIPS, J. C. *ibid.* **147**, 469–77 (1966).
108. HILSCH, R. and POHL, R. W. *Z. Physik* **59**, 812–17 (1930).
109. PHILIPP, H. R. and EHRENREICH, H. *Phys. Rev.* **131**, 2016–22 (1963).
110. PHILLIPS, J. C. *Phys. Rev.* **136**A, 1705–13 (1964).
111. PHILLIPS, J. C. *Phys. Rev.* **136**A, 1714–21 (1964).
112. ONADERA, Y., OKAZAKI, M. and INUI, T. *J. Phys. Soc. Japan*, **21**, 816 (1966); *ibid.* **21**, 2229–38 (1966).
113. PHILLIPS, J. C., *Phys. Rev.* **136**A, 1721–7 (1964).
114. CZYZAK, S. J., BAKER, W. M., CRANE, R. C. and HOWE, J. B. *J. Opt. Soc. Am.* **47**, 240–3 (1957).
115. BEUN, J. A. and GOLDSMITH, G. J. *Helv. Phys. Acta* **33**, 508–13 (1960).
116. THOMAS, D. G. and HOPFIELD, J. J. *Phys. Rev.* **116**, 573–82 (1959).
117. DIMMOCK, J. O. and WHEELER, R. G. *J. Appl. Phys.* **32**, 2271–7 (1961).
118. REYNOLDS, D. C., LITTON, C. W. and COLLINS, T. C. *Phys. Stat. Sol.* **9**, 645–84 (1965), and **12**, 3–55 (1965).
119. BIRMAN, J. L. *Phys. Rev.* **115**, 1493–1505 (1959).
120. CARDONA, M. and HARBEKE, G. *Phys. Rev.* **137**A, 1467–76 (1965).
121. CARDONA, M., WEINSTEIN, M. and WOLFF, G. A. *Phys. Rev.* **140**A, 633–7 (1965).
122. CARDONA, M. *Solid State Communications* **1**, 109–15 (1963).
123. BALKANSKI, M. and PETROFF, Y. *Proc. Int. Conf. on Semiconductor Physics*, Paris, 1964, pp. 245–50.
124. COLLINS, T. C., EUWEMA, R. N. and DEWITT, J. S. *Proc. Int. Conf. on Semiconductor Physics*, Kyoto, 1966, pp. 15–19.
125. BERGSTRESSER, T. K. and COHEN, M. L. *Physics Letters* **23**, 8–9 (1966).
126. RÖSSLER, U. and LIETZ, M. *Phys. Stat. Sol.* **17**, 597–604 (1966).
127. DRABBLE, J. A. The physical properties of single crystal bismuth telluride, *Progress in Semiconductors* **7**, 45–98 (1963).
128. AUSTIN, I. G. *Proc. Phys. Soc.* **72**, 545–52 (1958).
129. OLECHNA, D. J. and KNOX, R. S. *Phys. Rev.* **140**A, 986–93 (1965).
130. REITZ, J. R. *Phys. Rev.* **105**, 1233–40 (1957).
131. STUKE, J. and KELLER, H. *Phys. Stat. Sol.* **7**, 189–203 (1964).
132. SANDROCK, R. and TREUSCH, J. *Solid State Communications* **3**, 361–3 (1965); *Phys. Stat. Sol.* **16**, 487–97 (1966).
133. HAERING, R. R. and MROZOWSKI, S. The band structure and electronic properties of graphite crystals, *Progress in Semiconductors* **5**, 273–316 (1960).
134. TAFT, E. A. and PHILIPP, H. R. *Phys. Rev.* **138**A, 197–202 (1965).
135. BASSANI, F. and PASTORI, G. *Nuovo Cimento* **50B**, 95 (1967).
136. DEVORE, J. R. *J. Opt. Soc. Am.* **41**, 416–19 (1951).
137. TAUC, J. and ABRAHAM, A. *Czech. J. Phys.* **15**, 730–9 (1965).
138. PAUL, W. and BROOKS, H. Effect of pressure on the properties of germanium and silicon, *Progress in Semiconductors* **7**, 135–238 (1963).
139. HERMAN, F. *Proc. Int. Conf. on Semiconductor Physics*, Prague, 1960, pp. 20–8.
140. BASSANI, F. and BRUST, D. *Phys. Rev.* **131**, 1524–9 (1963).

141. BRAUNSTEIN, R., MOORE, A. R. and HERMAN, F. *Phys. Rev.* **109**, 695–710 (1958).
142. ZALLEN, R., PAUL, W. and TAUC, J. *Bull. Am. Phys. Soc.* **7**, 185 (1962).
143. TAUC, J. *Proc. Int. Conf. on Semiconductor Physics*, Exeter, 1962, pp. 333–44.
144. PAUL, W., DEMEIS, M. and FINEGOLD, L. X. *Proc. Int. Conf. on Semiconductor Physics*, Exeter, 1962, pp. 712–21.
145. CARDONA, M. *J. Appl. Phys.* **32**, 2151–5 (1961).
146. BRUST, D. *Phys. Rev.* **139**A, 489–500 (1965).
147. BERGSTRESSER, T. K., COHEN, M. L. and WILLIAMS, E. W. *Phys. Rev. Letters* **15**, 662–4 (1965).
148. WOOLLEY, J. C., THOMPSON, A. G. and RUBENSTEIN, M. *Phys. Rev. Letters* **15**, 670–2 (1965).
149. SHAKLEE, K. L., CARDONA, M. and POLLAK, F. H. *Phys. Rev. Letters* **16**, 48–50 (1966).
150. ENGELER, W. E., GARFINKEL, M., TIEMANN, J. J. and FRITZSCHE, H. *Proc. Int. Coll. on Optical Properties and Electronic Structure of Metals*, Paris, 1965, pp. 189–95.
151. GERHARDT, U. *Phys. Rev. Letters* **15**, 401–3 (1965).
152. HENSEL, J. C. and FEHER, G. *Phys. Rev.* **129**, pp. 1041–62 (1963).
153. FRENKEL, J. *Phys. Rev.* **37**, 17–44, 1276–94 (1931).
154. WANNIER, G. *Phys. Rev.* **52**, 191–7 (1937).
155. ELLIOTT, R. J. *Phys. Rev.* **106**, 1384–9 (1957).
156. MACFARLANE, G. G., MACLEAN, T. P., QUARRINGTON, J. E. and ROBERTS, V. *J. Phys. Chem. Solids* **8**, 388–92 (1959).
157. STURGE, M. D. *Phys. Rev.* **127**, 768–73 (1962).
158. BASSANI, F., GREENAWAY, D. L. and FISCHER G. *Proc. Int. Conf. on Semiconductor Physics*, Paris, 1964, pp. 51–6.
159. GROSS, E. F. *Suppl. Nuovo Cimento* **3**, 672–701 (1956).
160. NIKITINE, S. Exciton spectra in semiconductors and ionic compounds, *Progress in Semiconductors* **6**, 233–322 (1962).
161. DUKE, C. B. and SEGALL, B. *Phys. Rev. Letters* **17**, 19–22 (1966).
162. VELICKÝ, B. and SAK, J. *Phys. Stat. Sol.* **16**, 147–57 (1966).
163. PHILLIPS, J. C. *Phys. Rev. Letters* **12**, 142–5 (1964).
164. PHILLIPS, J. C. *Phys. Rev. Letters* **12**, 447–9 (1964).
165. FANO, U. *Phys. Rev.* **124**, 1866–78 (1961).
166. TOYAZAWA, Y., INOUE, M., INUI, T., OKAZAKI, M. and HANAMURA, E. *Proc. Int. Conf. on Semiconductor Physics*, Kyoto, 1966, pp. 133–8; *J. Phys. Soc. Japan* **21**, 1850–1 (1966).
167. MACFARLANE, G. G. and ROBERTS, V. *Phys. Rev.* **98**, 1865–6 (1955).
168. HOPFIELD, J. J. and THOMAS, D. G. *Phys. Rev.* **132**, 563–72 (1963).
169. CARDONA, M. and HARBEKE, G. *Phys. Rev. Letters* **8**, 90–1 (1962).
170. GREENAWAY, D. L. and CARDONA, M. *Proc. Int. Conf. on Semiconductor Physics*, Exeter, 1962, pp. 666–72.
171. HARBEKE, G. *Z. f. Naturforsch.* **19a**, 548–52 (1964).
172. PHILLIPS, J. C. *Phys. Rev. Letters* **10**, 329–32 (1963).
173. CLARK, C. D., DEAN, P. J. and HARRIS, P. V. *Proc. Roy. Soc.* **277**, 312–29 (1964).
174. PHILLIPS, J. C. *Phys. Rev.* **139**A, 1291–4 (1965).
175. ROBERTS, R. A., ROESSLER, D. M. and WALKER, W. C. *Phys. Rev. Letters* **17**, 302–4 (1966).
176. JAIN, K. P. *Phys. Rev.* **139**A, 544–59 (1965).

177. Greenaway, D. L. and Nitsche, R. *J. Phys. Chem. Solids* **26**, 1445–58 (1965).
178. Bachmann, K., Kneubühl, F. K. and Tegtmeier, A. *Helv. Phys. Acta* **36**, 804–6 (1963).
179. Greenaway, D. L. and Harbeke, G. *Proc. Int. Conf. on Semiconductor Physics*, Kyoto, 1966, pp. 151–5.
180. Franz, W. *Z. Naturforsch.* **13a**, 484–9 (1958).
181. Keldysh, W. L. *Zh. Eksperim, i. Teor. Fiz.* **34**, 1138–41 (1958) (English transl.: *Soviet Phys.-JETP* **7**, 788–90 (1958)).
182. Moss, T. S. *J. Appl. Phys.* **32**, 2136–9 (1962).
183. Callaway, J. *Phys. Rev.* **130**, 549–53 (1963); **134**A, 998–1000 (1964).
184. Tharmalingam, K. *Phys. Rev.* **130**, 2204–6 (1963).
185. Chester, M. and Fritsche, L. *Phys. Rev.* **139A**, 518–25 (1965).
186. Penchina, C. M. *Phys. Rev.* **138**A, 924–6 (1965).
187. Frova, A. and Handler, P. *Appl. Phys. Letters* **5**, 11–13 (1964).
188. Frova, A. and Handler, P. *Phys. Rev.* **137**A 1857–61 (1965).
189. Chester, M. and Wendland, P. *Phys. Rev. Letters* **13**, 193–5 (1964).
190. Frova, A. and Handler, P. *Phys. Rev. Letters* **14**, 178–80 (1965).
191. Seraphin, B. O. *Proc. Int. Conf. on Semiconductor Physics,* Paris, 1964, pp. 165–70.
192. Seraphin, B. O. and Bottka, N. *Phys. Rev.* **145**, 628–36 (1966); and Seraphin, B. O. *Phys. Rev.* **140A**, 1716–25 (1965).
193. Seraphin, B. O. *Proc. Phys. Soc.* **87**, 239–44 (1966).
194. Seraphin, B. O. and Bottka, N. *Phys. Rev.* **139**A, 560–5 (1965).
195. Phillips, J. C. *Phys. Rev.* **146**, 584–9 (1966).
196. Phillips, J. C. and Seraphin, B. O. *Phys. Rev. Letters* **15**, 107–10 (1965).
197. Aspnes, D. E. *Phys. Rev.* **147**, 554–66 (1966).
198. Hamakawa, Y., Germano, F. and Handler, P. *Proc. Int. Conf. on Semiconductor Physics,* Kyoto, 1966, pp. 111–17.
199. Spicer, W. E. and Simon, R. E. *J. Phys. Chem. Solids* **23**, 1817–20 (1962).
200. Gobeli, G. W., Allen, F. G. and Kane, E. O. *Phys. Rev. Letters* **12**, 94–7 (1964).
201. Allen, F. G. and Gobeli, G. W. *Phys. Rev.* **127**, 141–9 (1962).
202. Cohen, M. L. and Phillips, J. C. *Phys. Rev.* **139**A, 912–20 (1965).
203. Kane, E. O. *Proc. Int. Conf. on Semiconductor Physics,* Kyoto, 1966, pp. 37–45.
204. Kindig, N. B. and Spicer, W. E. *Phys. Rev.* **138A**, 561–76 (1965).
205. Spicer, W. E. *Phys. Rev. Letters* **11**, 243–5 (1963).
206. Spicer, W. E. and Lapeyre, G. J. *Phys. Rev.* **139**A, 565–9 (1965).
207. Spicer, W. E. *Physics Letters* **20**, 325–6 (1966).
208. Raether, H. Solid state excitations by electrons, *Springer Tracts in Modern Physics* **38**, 84–157 (1965).
209. Creuzburg, M. and Raether, H. *Solid State Communications,* **2,** 175–7 (1964).
210. Zeppenfeld, K. and Raether, H. Z. *Phys.* **193**, 471–8 (1966).
211. LaVilla, R. and Mendlowitz, H. *Phys. Rev. Letters* **9,** 149–50 (1962).

INDEX